PEER REVIEW AND DESIGN COMPETITION

IN THE NNSA NATIONAL SECURITY LABORATORIES

Committee on Peer Review and Design Competition
Related to Nuclear Weapons

Division on Engineering and Physical Sciences

The National Academies of
SCIENCES · ENGINEERING · MEDICINE

THE NATIONAL ACADEMIES PRESS
Washington, DC
www.nap.edu

THE NATIONAL ACADEMIES PRESS **500 Fifth Street, NW** **Washington, DC 20001**

This study was supported by Contract No. DE-PI0000010/DE-DT0006061 between the National Academy of Sciences and the U.S. Department of Energy. Any opinions, findings, conclusions, or recommendations expressed in this publication do not necessarily reflect the views of the agency that provided support for the project.

International Standard Book Number-13: 978-0-309-37843-7
International Standard Book Number-10: 0-309-37843-5

Copies of this report are available from

Laboratory Assessments Board
Division on Engineering and Physical Sciences
National Research Council
500 Fifth Street, NW
Washington, DC 20001

Additional copies of this report are available for sale from the National Academies Press, 500 Fifth Street, NW, Keck 360, Washington, DC 20001; (800) 624-6242 or (202) 334-3313; http://www.nap.edu.

Copyright 2015 by the National Academy of Sciences. All rights reserved.

Printed in the United States of America

Suggested citation: National Academies of Sciences, Engineering, and Medicine. 2015. *Peer Review and Design Competition in the NNSA National Security Laboratories.* Washington, DC: The National Academies Press.

The National Academies of

SCIENCES · ENGINEERING · MEDICINE

The **National Academy of Sciences** was established in 1863 by an Act of Congress, signed by President Lincoln, as a private, nongovernmental institution to advise the nation on issues related to science and technology. Members are elected by their peers for outstanding contributions to research. Dr. Ralph J. Cicerone is president.

The **National Academy of Engineering** was established in 1964 under the charter of the National Academy of Sciences to bring the practices of engineering to advising the nation. Members are elected by their peers for extraordinary contributions to engineering. Dr. C. D. Mote, Jr., is president.

The **National Academy of Medicine** (formerly the Institute of Medicine) was established in 1970 under the charter of the National Academy of Sciences to advise the nation on medical and health issues. Members are elected by their peers for distinguished contributions to medicine and health. Dr. Victor J. Dzau is president.

The three Academies work together as the **National Academies of Sciences, Engineering, and Medicine** to provide independent, objective analysis and advice to the nation and conduct other activities to solve complex problems and inform public policy decisions. The Academies also encourage education and research, recognize outstanding contributions to knowledge, and increase public understanding in matters of science, engineering, and medicine.

Learn more about the **National Academies of Sciences, Engineering, and Medicine** at **www.national-academies.org.**

COMMITTEE ON PEER REVIEW AND DESIGN COMPETITION RELATED TO NUCLEAR WEAPONS

PAUL S. PEERCY, University of Wisconsin, Madison, *Co-Chair*
JILL P. DAHLBURG, Naval Research Laboratory, *Co-Chair*
JOHN F. AHEARNE, Sigma Xi, The Scientific Research Society
MICHAEL R. ANASTASIO, Los Alamos National Laboratory (retired)
CHRISTINA A. BACK, General Atomics
JOHN M. CORNWALL, University of California, Los Angeles
PAUL A. FLEURY, Yale University
DAVID HAMMER, Cornell University
CHERRY A. MURRAY, Harvard University
ROBERT E. NICKELL, Applied Science & Technology[1]
DAVID OVERSKEI, Decision Factors, Inc.
K. LEE PEDDICORD, Texas A&M University
ROBERT SELDEN, Independent Consultant
STEVEN J. ZINKLE, University of Tennessee, Knoxville

Staff

SCOTT WEIDMAN, Responsible Staff Officer
GREG EYRING, Senior Program Officer
RICHARD ROWBERG, Deputy Director, Division of Engineering and Physical Sciences
RODNEY HOWARD, Administrative Assistant

[1] Resigned November 7, 2014.

LABORATORY ASSESSMENTS BOARD

JOHN W. LYONS, U.S. Army Research Laboratory (retired), *Chair*
ROSS B. COROTIS, University of Colorado at Boulder
PAUL A. FLEURY, Yale University
C. WILLIAM GEAR, NEC Research Institute, Inc. (retired)
WESLEY L. HARRIS, Massachusetts Institute of Technology
JENNIE S. HWANG, H-Technologies Group, Inc.
W. CARL LINEBERGER, University of Colorado at Boulder
C. KUMAR N. PATEL, Pranalytica, Inc.
ELSA REICHMANIS, Georgia Institute of Technology
LYLE H. SCHWARTZ, University of South Florida

Staff

JAMES P. McGEE, Director
LIZA HAMILTON, Associate Program Officer
EVA LABRE, Administrative Coordinator
ARUL MOZHI, Senior Program Officer
ANDREA SHELTON, Administrative Assistant

Preface

In the FY2013 National Defense Authorization Act, Congress directed the administrator of the Department of Energy's (DOE's) National Nuclear Security Administration (NNSA) to reach an agreement with the National Academies of Sciences, Engineering, and Medicine[1] to "conduct a study of peer review and design competition related to nuclear weapons" at its national security laboratories (see Appendix C for the legislation language). The NNSA laboratories involved are Los Alamos National Laboratory (LANL), Lawrence Livermore National Laboratory (LLNL), and Sandia National Laboratories (SNL). The full Statement of Task is discussed in Chapter 1.

The Administrator commissioned the Academies to conduct this study on peer review and design competition in the context of nuclear weapons. In response, the Academies formed the Committee on Peer Review and Design Competition Related to Nuclear Weapons, which began work on June 10, 2014. For the backgrounds of committee members, see Appendix A. The committee had the benefit of presentations from a number of individuals with knowledge and experience related to its tasks; the agendas of the committee's public meeting sessions are listed in Appendix B.

As the committee was conducting this study, it was mindful of other

[1] Effective July 1, 2015, the institution is called the National Academies of Sciences, Engineering, and Medicine. References in this report to the National Research Council (NRC) are used in a historical context to refer to activities before July 1.

ongoing studies with overlapping mandates.[2] In particular, another Academies study, *Aligning the Governance Structure of the NNSA Laboratories to Meet 21st Century National Security Challenges*, was mandated in the same legislation and dealt in part with peer review issues at the NNSA laboratories; the staff of the two studies coordinated with each other

We wish to thank all of the committee members for their dedication in producing this report in a short time. Also, we particularly want to thank Michael Bernardin of LANL, Mike Dunning of LLNL, Gary Sanders and Ron Hartwig of SNL, and Robert Hanrahan of NNSA for their time and efforts in organizing the presentations that the committee received during its laboratory visits, for attending and contributing to the committee's public meetings, and for responding in a timely way to the committee's requests for additional information. The committee is well aware of the burdens imposed by the laboratories' support of outside groups seeking to review them—including this committee—and hopes that the laboratories will feel that the time spent was worthwhile. Finally, the outside reviewers and the National Academies monitor provided insightful comments that improved the quality of the report. A sincere thank you is due to the Academies staff: Scott Weidman, Dick Rowberg, Greg Eyring, and Rodney Howard.

Paul S. Peercy and Jill P. Dahlburg, *Co-Chairs*
Committee on Peer Review and Design
Competition Related to Nuclear Weapons

[2] Three other related, congressionally mandated studies were under way or completed as this study progressed: (1) Congressional Advisory Panel on the Governance of the Nuclear Security Enterprise, *A New Foundation for the Nuclear Enterprise*, November 2014, available at http://knoxblogs.com/atomiccity/2014/12/11/reforming-nuclear-security-enterprise/ (its final report); (2) Commission to Review the Effectiveness of the National Energy Laboratories, *Interim Report of the Commission to Review the Effectiveness of the National Energy Laboratories*, February 27, 2015, http://energy.gov/labcommission/downloads/interim-report-commission-review-effectiveness-national-energy-laboratories; and (3) NRC, *Aligning the Governance Structure of the NNSA Laboratories to Meet 21st Century National Security Challenges*, The National Academies Press, Washington, D.C., 2015. In addition, there is a Secretary of Energy Advisory Board Task Force on DOE National Laboratories; see http://energy.gov/seab/secretary-energy-advisory-board-seab-task-force-doe-national-laboratories.

Acknowledgment of Reviewers

This report has been reviewed in draft form by individuals chosen for their diverse perspectives and technical expertise, in accordance with procedures approved by the Report Review Committee. The purpose of this independent review is to provide candid and critical comments that will assist the institution in making its published report as sound as possible and to ensure that the report meets institutional standards for objectivity, evidence, and responsiveness to the study charge. The review comments and draft manuscript remain confidential to protect the integrity of the deliberative process. We wish to thank the following individuals for their review of this report:

Everett H. Beckner, Independent Consultant, Santa Fe, New Mexico,
Philip E. Coyle, Private Consultant, Sacramento, California,
Thomas O. Hunter, Sandia National Laboratories (retired),
Raymond Jeanloz, University of California, Berkeley,
John Kammerdiener, Independent Consultant, Marble Falls, Texas,
Steven Lamoreaux, Yale University,
Charles V. Shank, Howard Hughes Medical Institute,
Merri Wood-Schultz, Los Alamos National Laboratory, and
Joan B. Woodard, EDT Consulting, LLC.

Although the reviewers listed above have provided many constructive comments and suggestions, they were not asked to endorse the conclusions or recommendations, nor did they see the final draft of the report before its release. The review of this report was overseen by Stephen Robinson, University of Wisconsin, Madison, who was responsible for making certain that an independent examination of this report was carried out in accordance with institutional procedures and that all review comments were carefully considered. Responsibility for the final content of this report rests entirely with the authoring committee and the institution.

Contents

Summary

During the Cold War, the United States designed, built, tested, and deployed numerous nuclear warheads of various designs. The results of nuclear tests of the nuclear explosive packages (NEPs) of these warheads provided the ultimate validation of the design procedures, weapon design codes, and designer judgment. Formal design competitions between teams from Los Alamos National Laboratory (LANL) and Lawrence Livermore National Laboratory (LLNL)—each supported by separate branches of Sandia National Laboratories (SNL) for the non-nuclear components and integration with the delivery system—were routinely held. Extensive peer reviews of the types in use today were less frequent and less formal during that era, primarily because of the availability of nuclear explosion testing.

Following the moratorium on nuclear explosion testing that began in 1992, all three National Nuclear Security Administration (NNSA) laboratories intensified their use of technical questioning, collaborative and competitive reviews, tests of components and subsystems, and modeling and simulation to regularly check their work and ensure that a range of perspectives is brought to bear so as to improve quality and uncover any problems. The laboratories have also strengthened their technical evaluation and peer review processes both to ensure the safety, security, and effectiveness of the nuclear stockpile and to help maintain associated science and engineering design and innovation capabilities in the laboratories. Although not a substitute for weapons tests, peer review has become an increasingly important practice at the three NNSA laboratories as a means of mitigating the risks that the nation's nuclear weapons will fail to perform as expected if needed.

After an assessment of peer review and design competition, the committee reached four conclusions:

Conclusion 1.1: In the main, peer review processes used by all three NNSA laboratories are healthy and robust, providing benefits such as increasing confidence in weapon assessment and certification, improving our understanding of weapons physics, addressing weapon aging issues, and identifying lower-cost approaches to Life-Extension Programs.

Conclusion 1.2: Incentives for peer review at the NNSA laboratories are abundantly evident.

- **Peer review reduces the risk of overlooking a technical option, of relying on suboptimal data or methods, or of simply making an embarrassing mistake, and thereby supports the ingrained culture at all three laboratories, which sets a high standard for quality.**
- **Peer review is visibly valued by laboratory management.**

Conclusion 1.3: SNL and the NEP design laboratories (LANL and LLNL) have taken somewhat different approaches to peer review, owing in large part to SNL's ability to test non-nuclear components and systems.

- **With only archival nuclear explosion test data available, LANL and LLNL rely on vigorous, deep-dive reviews by true competitive peers and other subject-matter experts to critique the results of calculations and subcritical experiments relating to NEP performance.**
- **With testing data available to verify the performance of components and systems and to validate modeling and simulation tools, peer review at SNL is driven more by the need to assure cost-effective performance of stockpile hardware under all anticipated conditions and by budget pressures to reduce the number of expensive tests.**

Conclusion 1.4: All three NNSA laboratories have opportunities to improve their processes for peer review:

- **With the exception of major reviews associated with the Annual Assessment Report or Life-Extension Programs, LLNL and LANL lack written guidance for conducting peer reviews to determine in general when a review is needed, how the review is to be conducted, who should participate in the review, or how to address review findings.**
- **SNL has developed useful written guidance for conducting peer reviews; however, during its visit to SNL and its probing of the presentations made, the committee determined that SNL could profitably make greater use of outside experts in its peer reviews, as called for in its written guidance.**

These conclusions led the committee to the following recommendation:

Recommendation 1: The nuclear weapons laboratories should improve their peer review processes in the following ways:

- **Los Alamos National Laboratory and Lawrence Livermore National Laboratory should ensure they have short, written guidance for a graded approach to peer review, the rigor of which is appropriate to the stage of work and range of technical activities being reviewed.**
- **Sandia National Laboratories should strengthen and broaden its use of outside experts on its peer review teams, as articulated in written guidance that Sandia recently finalized.**

In the area of design competition, the committee reached one conclusion and one recommendation:

Conclusion 2: The innovations produced by design competitions during the Cold War, as well as the increased confidence in the safety and reliability of stockpile weapons resulting from current assessment processes such as the Independent Nuclear Weapons Assessment Process (INWAP), illustrate the value of having independent teams, using different approaches and methods, addressing common problems.

Recommendation 2: Los Alamos National Laboratory and Lawrence Livermore National Laboratory should continue to maintain independent design capabilities, using different approaches and methods, to enable independent peer review of critical technical issues. Sandia National Laboratories should likewise carry out, for high-priority issues, competitive designs with independent teams that

use different approaches, followed by peer reviews of components, subsystems, and full systems, as discussed in Recommendation 1.

The Reliable Replacement Warhead (RRW) design study produced innovative designs by competing teams at LANL/SNL-New Mexico and LLNL/SNL-California.[1] However, the manner in which the study was conducted led to deep resentments at the laboratories.

Conclusion 3: Although the RRW design study succeeded in producing innovative weapon designs by the competing teams, its value was reduced because technical experts from the competing laboratories were not given the opportunity to critique one another's ideas through interlaboratory peer review or to address criticisms at the science and engineering level before the final designs were formally presented to NNSA and potential end users.

Recommendation 3: To guide future design studies and design competitions, the National Nuclear Security Administration (NNSA) should provide a formal written statement articulating the design requirements and objectives, along with the selection criteria, in advance of any authorized work. NNSA should also ensure that interlaboratory peer review takes place and that competitors have an opportunity to address criticisms at the science and engineering level before the results are formally presented to stakeholders outside NNSA.

Finally, the committee is deeply concerned about the state of design competition at all three laboratories. There have been no full[2] design competitions for NEPs since the 1992 moratorium on the testing of nuclear explosions. The Department of Defense (DOD) has not asked for any fundamentally new warhead designs, and for a considerable time Congress limited work on new designs.

---

[1] The Reliable Replacement Warhead design study was intended to generate competitive designs for a highly reliable warhead with enhanced surety. In addition, the competition aimed for a design that could be manufactured relatively easily with currently available materials, eliminating some potentially hazardous materials that had been used previously. Finally, it was necessary that the weapon as designed could be certified without nuclear explosion testing.

[2] By "full design competition" the committee means competitions that integrate the full end-to-end design process, including design of the integrated NEP—from conceiving of a novel design to address a threat, through production of an engineering prototype, a step that provides essential feedback about the practicality of a design. Full design competition would not include nuclear explosion testing.

Design competitions, and the subsequent testing of components, subsystems, and systems (within the limits of national policy and agreements) are critical to developing the next generation of nuclear weapons designers with expertise that goes beyond analysis and modeling. Although it was considered too expensive for every design competition to result in the production of a prototype during the Cold War, those that did provided the feedback that designers needed to stay at the cutting edge. The number of the NEP laboratories' science and engineering personnel with hands-on experience in nuclear weapons design and nuclear explosion testing continues to decrease and will reach zero in the next decade or so. Once this experience is lost, it could limit the nation's strategic options, and it will be difficult to reestablish.

Looking to the future, maintaining nuclear weapons design skills at the NEP laboratories—as well as production skills within the NNSA complex—is essential to achieve three objectives:

1. Maintaining a credible nuclear deterrent workforce that is capable of designing and building weapons to meet evolving threats;
2. Understanding the status and direction of foreign nuclear weapon programs, and thus strengthening the nonproliferation regime; and,
3. Determining the best and most cost-effective approaches to resolving problems that arise during stockpile weapon surveillance and life extension programs.

To avoid the potential of losing a capability that could be essential for responding to evolving threats, the NNSA complex needs a means of exercising, on a regular and ongoing basis, the full suite of nuclear weapons design, development, and engineering capabilities through true design competitions. Thus, the committee makes the following conclusion and recommendation.

Conclusion 4: In contrast to the robust state of peer review at the NNSA laboratories, the state of design competition is not robust.

- **There have been no full NEP design competitions since the 1992 nuclear explosion testing moratorium. Recent design studies have been good analysis and modeling exercises, but they did not result in actual engineering and fabrication of components and systems; thus, they did not exercise the complete set of skills required in the NNSA complex to design nuclear weapons that would be an effective deterrent, nor was the credibility of any design assessed by fabricating a device or by non-nuclear testing.**

- **At SNL, the need to continually replace aging or obsolete non-nuclear components in stockpile weapons, as well as the large Life-Extension Programs for the W76 and B61, have indeed exercised designers' skills. However, these exercises do not stimulate the full creativity and innovation that result from a true blank slate design competition that includes engineering and building a prototype.**

Recommendation 4: In order to exercise the full set of design skills necessary for an effective nuclear deterrent, the National Nuclear Security Administration should develop and propose the first in what the committee envisions as a series of design competitions that include designing, engineering, building, and non-nuclear testing of a prototype. The non-nuclear components produced by Sandia should be integrated into the design and fabrication of the prototype. This should be done with the clear understanding that this prototype would not enter the stockpile.

Implementation of the committee's four recommendations would help develop and maintain the most important asset—a competent workforce with demonstrated skills and judgment—and instill confidence by all stakeholders (including adversaries) in the ability of this workforce to maintain the nuclear deterrent.

1

Introduction and Charge

The National Nuclear Security Administration (NNSA) is responsible for providing and maintaining the capabilities necessary to sustain a safe, secure, and reliable nuclear weapons stockpile for the nation and its allies. Major responsibility for meeting the NNSA missions falls to the three NNSA laboratories: Los Alamos National Laboratory (LANL), Lawrence Livermore National Laboratory (LLNL), and Sandia National Laboratories (SNL). The NNSA national security laboratories contribute to that goal by maintaining the skills and capabilities necessary for stewardship of a reliable nuclear stockpile and also by maintaining a high level of technical credibility, which is a component of the nuclear deterrent.

Since 1992 it has been U.S. policy not to conduct explosion tests of nuclear weapons.[1] The resulting technical challenges have been substantial. Whereas a nuclear test was in some sense the ultimate "peer review" of the performance of a particular NEP design, the cessation of nuclear testing necessitated a much greater reliance on both intralab and interlab expert peer review to identify potential problems with weapon designs and define the solution space.

[1] On September 24, 1996, the United States signed the Comprehensive Nuclear Test Ban Treaty, although it had been observing a nuclear explosion testing moratorium since 1992. The U.S. Senate has not ratified the treaty, but the United States has observed the treaty and fulfilled its obligations under the treaty.

COMMITTEE CHARGE

The statement of task to which this report responds reads as follows:

Assess the following:

- The quality and effectiveness of peer review of designs, development plans, engineering and scientific activities, and priorities related to both nuclear and non-nuclear aspects of nuclear weapons;
- Incentives for effective peer review;
- The potential effectiveness, efficiency, and cost of alternative methods of conducting peer review and design competition related to both nuclear and non-nuclear aspects of nuclear weapons, as compared to current methods;
- Known instances where current peer review practices and design competition succeeded or failed in finding problems or potential problems; and
- How peer review practices related to both nuclear and non-nuclear aspects of nuclear weapons should be adjusted as the three NNSA laboratories transition to a broader national security mission.

The last task seeks to explore how the evolving mission of the NNSA laboratories—from an exclusive focus on nuclear weapons in the 1950s to a broader national security mission today—might impact peer review processes at the laboratories that relate to nuclear weapons.

The committee has understood that the "effectiveness" of peer review referred to in its charge is the effectiveness that can be achieved under the current conditions in which the nuclear weapons program now operates—that is, without nuclear explosion testing.

HISTORICAL CONTEXT

Founded in 1943, Los Alamos was the first U.S. nuclear weapons laboratory, followed by Sandia Laboratory in 1949. Sandia Laboratory was originally the Z-Division of Los Alamos, which designed and developed the non-nuclear components of the nuclear weapons and was later relocated to Kirtland Air Force Base in Albuquerque, New Mexico. Lawrence Livermore Laboratory was formed in 1952. In 1956, a branch of Sandia was also established in Livermore, California, to design and develop the non-nuclear components of nuclear weapons in support of Lawrence Livermore-designed nuclear explosive packages (NEPs). Following multiple organizational transitions subsequent to their formation, the three nuclear weapons laboratories became today's three NNSA national security laboratories: LANL, LLNL, and SNL. Their role within the national nuclear security enterprise is illustrated on pages 4 and 5 of the recent "Augustine-Mies" report.[2]

Throughout the Cold War, the two distinctly separate system design teams[3]—the LLNL/SNL-California design collaboration team and the LANL/SNL-New Mexico design collaboration team—competed intensely to win the right to design each new weapon system.[4]

During the Cold War, the ultimate validation of NEP design procedures, design codes, and designer judgment was provided by a nuclear test or a series of tests. Design competitions during the Cold War—and the associated design, development, production, and testing of new nuclear weapons—enabled the nuclear weapons complex to respond to evolving strategic threats with a strong and reliable nuclear deterrent. The nuclear explosion testing moratorium implemented in 1992 resulted in a fundamental shift in the approach for maintaining the safety, security, and effectiveness of the U.S. nuclear weapons stockpile. Following the cessation of nuclear explosion testing and of new weapon development in 1992, comprehensive design competition for NEPs in which the designs were validated with manufactured prototypes essentially ceased. The result is that today the full design expertise at LANL, LLNL, and SNL is no longer directly exercised.

The Science-Based Stockpile Stewardship Program[5] was introduced by the Department of Energy (DOE) in the mid-1990s to strengthen the foundations of knowledge on which the laboratories' work relies, especially to enable the laboratories to assess the safety, security, and effectiveness of the stockpile without nuclear testing. This included, in particular, a major effort to extend the capabilities of computer modeling and simulation to offset partially the loss of the information stream that had been provided by nuclear tests.

The committee strongly supports the position that it is a core responsibility of the three NNSA national security laboratories to sustain the essential science and engineering (S&E) capabilities that enabled the United States to successfully develop and manufacture its current stockpile, as these capabilities are essential for extending the life of the stock-

[2] Congressional Advisory Panel on the Governance of the Nuclear Security Enterprise, *A New Foundation for the Nuclear Enterprise: Report of the Congressional Advisory Panel on the Governance of the Nuclear Security Enterprise*, November 2014.

[3] The separate system design teams in Livermore and Albuquerque relied on common Sandia component design groups in Albuquerque, e.g., one radar design group, one neutron generator design group.

[4] Though often the task of taking the design forward to detailed design was assigned to the team that originated the design, occasionally the design was assigned to the other laboratory team for "workload leveling."

[5] The name was later shortened to the Stockpile Stewardship Program, or SSP.

pile (as individual weapon components age and/or are replaced by substitute components), to evaluate intelligence about potential adversaries, and to address what-if questions that arise as threats evolve.

At present, some S&E experts at the laboratories who have direct experience with nuclear weapons design, testing, and subsequent manufacturing continue to be active in all three national security laboratories. However, with the passage of time, there will soon be no laboratory personnel with first-hand experience in NEP design or the practical insights gained from testing their designs. Such hands-on experience and the judgment that derives from it is necessary for stockpile stewardship and for addressing technical challenges of nuclear nonproliferation. Scientists and engineers are well aware that models and theory for highly complex physical processes are incomplete and must absolutely be validated by experiments. The tests of actual weapons before the test moratorium provided essential understanding of how different design choices affected actual performance, and those test results were sometimes strikingly different from the best predictions then available. By engaging in design, production, and nuclear explosion testing, weapons designers built up invaluable judgment about the interplay between predicted and actual performance.

In the absence of nuclear explosion testing, all three laboratories today apply technical questioning, collaborative and competitive reviews, various tests of subsystems, and modeling and simulation to regularly check work and ensure that a range of perspectives is brought to bear. These processes, along with data from improved computational and experimental facilities, help current and future S&E staff develop the insights and expertise that had previously been gained by experience with direct nuclear tests. The laboratories have also strengthened their technical evaluation and peer review processes to ensure the safety, security, and effectiveness of the nuclear stockpile and to help maintain the associated S&E design and innovation capabilities.

PEER REVIEW

Peer review has become an increasingly important practice at the three NNSA laboratories as a means of reducing the risk of misconceptions and errors slipping into their science and engineering activities. Rigorous peer review is essential to maintaining high standards of science, technology, and design and to sustaining confidence in the performance of an aging stockpile. More generally, peer review at the laboratories—as throughout the research enterprise worldwide—is recognized as a means of ensuring high-quality work products. It thus contributes to the vibrancy and technical credibility of the laboratories' S&E workforce, which, as mentioned at the beginning of this chapter, is of fundamental importance to the nation's nuclear deterrent capability.

All three laboratories recognize the need for multiple types of review, including the following:

- External peer review, involving persons actively engaged in equivalent work outside the laboratory who review a laboratory's work in a particular area;
- Internal peer review, involving persons from the same laboratory actively engaged in equivalent work (but not the work being reviewed) who review the laboratory's work in the given area;
- Subject-matter reviews, involving internal and external subject-matter experts, who review a laboratory's execution of, or capabilities in, particular areas of science or engineering; and
- Technical reviews of programs or projects, involving internal and external experts who may not have as full a range of expertise as the people performing the original work but who provide technical scrutiny and checking of selected aspects of the work.

Examples of Peer Review

Because of the wide variation in how "review" and "peer review" are understood, the committee found it useful to develop a taxonomy that describes the primary types of review and structured competition activities that are valuable for strengthening S&E quality at the laboratories:

- *Traditional peer review.* A review by an expert or a body of experts who are independent of the activity under review but who could have carried out that activity.
- *Independent weapon assessment.* Teams from each laboratory independently assess a weapon type[6] for the Annual Assessment Report (AAR) prepared by each laboratory for the use of the laboratory director in preparing the annual letter to the President assessing the readiness of that weapon. Similarly, independent teams review evidence before closing a significant finding investigation (SFI), which can arise from regular surveillance of

[6] To "assess" a weapon type in the stockpile means to analyze all existing relevant data, including data from surveillance, experiments, and previous nuclear tests, to gauge whether there are any issues with its safety, security, and reliability. The assessment also involves computer simulations using the latest models of the weapon's performance consistent with the existing data.

stockpiled weapons,[7] and in reviewing plans for Life-Extension Programs (LEPs), in which all the weapons of a given type are refurbished or have aging parts replaced.

- *Red teaming.* A special category of peer review, in which the reviewers actively seek to find serious flaws in the work under review. Beginning in 2003, red team reviews were mandated by law as part of the AAR. For an AAR, a red team composed of representatives from all three weapons laboratories reviews the assessments submitted to the NEP design laboratory directors, and those directors take into account the red team reports in crafting their annual assessment letters.
- *Outside expert review.* A review of an activity by subject-matter experts in some or all of the technical components of the reviewed activity. Subject-matter experts typically come from outside the laboratories but have the appropriate clearances. An example would be the many reviews conducted by the JASON group. In the context of this report, a review by JASON would differ from the traditional peer review described above because the outside experts might have backgrounds in disciplines other than nuclear weapons. As a result, they are not peers in the sense of themselves being able to carry out the work being reviewed.

Metrics

The most common metric used by the nuclear weapons community is quantification of margins and uncertainties, or QMU. "QMU is a decision-support framework that provides a means for quantifying the laboratories' confidence that the critical stages of a nuclear weapon will operate as intended."[8] QMU systematically applies the output of the Stockpile Stewardship Program (aboveground non-nuclear and subcritical experiments, data from past underground nuclear tests, sophisticated modeling, and the expert judgment of weapons scientists) to the assessment of

[7] Patrick Garcia, LANL, "SFIs—Connections to Design & Peer Review," Presentation to the committee on September 24, 2014. Problems identified during weapon surveillance programs trigger SFIs, which involve an evaluation of potential impact and required response, and those study results undergo a management review. If management deems an SFI to be more than a one-off issue with minimal impact and an obvious response, an internal peer review is used if the impact or response is not completely clear and some independent validation is warranted, whereas an external peer review is used if the impact is viewed as potentially significant and design agency consensus is warranted.

[8] National Research Council, 2009, *Evaluation of Quantification of Margins and Uncertainties Methodology for Assessing and Certifying the Reliability of the Nuclear Stockpile*, The National Academies Press, Washington, D.C., p. 5.

the stockpile. Simply stated, Q = M/U, where Q is a measure of assured performance, M is the margin of uncertainty, and U is the uncertainty of M. The framework helps identify areas of risk with the greatest impact on performance and thus to set priorities for investment and testing. Because the application of QMU relies heavily on expert judgment, it relies strongly on peer review at all three laboratories.

DESIGN COMPETITION

Design competition is not a type of peer review. Rather, it refers to a process in which independent teams compete to design warheads that offer the best response to a specified set of goals and requirements. Whereas the primary goal of peer review is to check the quality of cutting-edge work, design competition fosters parallel efforts that vie to push frontiers. In this report the committee distinguishes two types of design competition:

- *Design studies.* These produce paper analyses and modeling and simulation results from competing teams but are not carried through to the production of a prototype. Perhaps the most complete example is the design study done for the Reliable Replacement Warhead (RRW),[9] as discussed in Chapter 3.
- *Full design competition.* Design competition as practiced during the Cold War exercised the full range of skills in the weapons complex: design, engineering, prototyping, testing, and production. The designers involved received feedback about the feasibility of their designs from the engineering process and from peers in the production facility, and later from actual nuclear tests. As used in this report, full design competition means a competition whose winning design would be carried through to a prototype device. The device would not be manufactured for the stockpile and would only be tested in a manner consistent with U.S. treaty obligations—that is, without nuclear yield.

Ultimately, the goal of reviews and competitions in the present day is to mitigate, to the extent possible in the absence of nuclear testing, the risk that the nation's nuclear weapons will fail to perform as expected if

[9] The RRW design study was intended to generate competitive designs for a highly reliable warhead with enhanced surety. In addition, the competition aimed for a design that could be manufactured relatively easily with currently available materials, eliminating some potentially hazardous materials that had been used previously. Finally, it was necessary that the weapon as designed could be certified without nuclear explosion testing.

needed. For example, reviews are intended to ensure that all significant technical risks have been identified in an LEP for a weapon, that a sound plan to mitigate those risks has been developed, and that the plan is being well implemented. And, limited-scope design competitions are embarked upon to explore a broader set of refurbishment options while at the same time providing an opportunity to exercise and transfer necessary skill sets to the next generation of nuclear weapons designers, in closer analogy to the successful modes of activity that were practiced during the Cold War.

In this report, the committee evaluates the efficacy of peer review and design competition in today's national nuclear security program and provides recommendations for how to ensure robust and reliable processes for peer review and design competition in the future.

CONDUCT OF THE STUDY

This study was mandated in Public Law 112-239, the National Defense Authorization Act for Fiscal Year 2013, Sec. 3144. It is sponsored by the National Nuclear Security Administration. The details of the legislation mandating the study are given in Appendix C. In response to the congressional mandate, the National Research Council[10] formed the Committee on Peer Review and Design Competition Related to Nuclear Weapons. The committee was formed so as to include people with experience in research and management in academe, national laboratories, and industry. Its members have various direct and indirect interactions with one or more of the NNSA laboratories in order for the committee to have insight into special circumstances associated with nuclear weapons science and engineering. All members hold security clearances at a level sufficient to access appropriate technical details. Biographical sketches of the committee members are included in Appendix A.

The study committee began its work with an open meeting on June 10, 2014, to receive background on the study and engage in discussions about the most important elements of peer review and design competition at the NNSA laboratories. The open session was preceded and followed by closed sessions in which the committee established a framework for how it would conduct its study. The study committee met in closed session the next day, June 11, 2014, to further develop its plans.

To collect information about peer review and design competition and to assess the effectiveness of current peer review and design competition practices, the study committee met subsequently at each of the three

[10] Effective July 1, 2015, the institution is called the National Academies of Sciences, Engineering, and Medicine. References in this report to the National Research Council (NRC) are used in a historical context to refer to activities before July 1.

laboratories to engage with a range of senior managers and staff. These meetings included presentations and discussions on topics contained in the charge. In each of these meetings, senior managers from all three laboratories were in attendance and participated in the discussions. The committee provided the laboratories in advance with specific questions it wished to investigate at these meetings and relied on each laboratory to arrange for an appropriate set of speakers and topics. Meetings at the laboratories were held in a classified setting, as needed, to allow for discussion of specific details. In the course of its information gathering, the committee also gained insight into current peer review practices in the Russian and U.K. nuclear enterprises—the former through a discussion with former LANL director Siegfried Hecker and the latter through a meeting with three experts associated with the U.K.'s atomic weapons establishment. Details of all information-gathering sessions at these meetings are given in Appendix B.

OUTLINE OF THE REPORT

The chapters that follow are organized chronologically, looking first at peer review and design competition during the period when nuclear explosion tests were still being conducted and new weapons developed, followed by a discussion of today's practices, and, finally, making recommendations to improve peer review and design competition at the laboratories going forward.

As discussed in Chapter 2, during the Cold War the United States designed, built, tested, and deployed numerous nuclear warheads of various designs. As noted above, the results of nuclear tests provided the ultimate validation of the NEP design procedures, weapon design codes, and designer judgment. Extensive peer reviews of the types used today were neither necessary nor practiced because the option of nuclear testing was available. That option notwithstanding, formal design competitions between teams from different laboratories were routinely held and led to important innovations in weapons designs. This chapter identifies key attributes of the laboratory practices during the Cold War that contributed to U.S. success in building and sustaining an effective nuclear arsenal and deterrent.

Chapter 3 presents the programs and practices for peer review and design competition currently used in the three national security laboratories.

As the NEP laboratories' experience base with weapons design, production, and testing fades, and as the international nuclear weapons landscape evolves, the NNSA laboratories face formidable challenges, discussed in Chapter 4. How they respond to these challenges is of vital importance for the nation. Effective peer review and design competition is part of the solution, as already recognized by NNSA and the laboratories, but a broader approach is needed. Chapter 4 provides conclusions and recommendations for strengthening current practices of peer review and design competition related to nuclear weapons and for maintaining a credible nuclear weapons design capability and an effective deterrent.

2

The Past: Before the 1992 Nuclear Explosion Testing Moratorium

There are two distinct periods in the history of the nuclear weapon laboratories: the Cold War period, during which nuclear explosion tests were conducted, and the period since 1992, when the nuclear explosion testing moratorium went into effect and new weapons development ceased. The vast majority of the warheads currently in the nuclear weapon stockpile were designed and tested during the Cold War period, which is the focus of this chapter.

The nuclear weapons era began in 1942. The Manhattan Project brought together many of the finest scientific minds in the United States and allied countries. This created a culture appreciative of intellectual excellence, which has continued to this day at all three laboratories.

The Manhattan Project began in academic settings, but in 1943 a weapons design laboratory was sited at Los Alamos, where the world's first nuclear devices were created within about 2 years. The Manhattan Project had a sense of urgency brought about by the war and by the perception that Germany was working on an atomic bomb. Those involved felt that a race was on and the consequences of Nazi Germany being first to possess a nuclear weapon were unacceptable—and indeed appalling.

The Z Division of Los Alamos Laboratory was created in 1945 to provide military liaison, ordnance engineering, and surveillance in storage sites and to manage non-nuclear testing. This organization was physically located on Sandia Base near Albuquerque, New Mexico, to facilitate better interactions with the military. Z Division also began the development of test ranges on Sandia Base and in California during this period.

It was soon recognized that management of all those functions from Los Alamos was inefficient and that local management with industrial connections could be more effective. As a result, Sandia Corporation was formed in 1949 to oversee staff and facilities on Sandia Base, and AT&T agreed to be the first industrial manager of the Sandia Corporation.

A second nuclear explosive package (NEP) design laboratory was established in Livermore, California, in 1952 to provide competition to Los Alamos. It provided intellectually stimulating and productive competition, which expanded the potential design space for nuclear weapons. The Lawrence Livermore Laboratory, in California, was chartered to work on designs that were deliberately different from, but competitive with, the designs being pursued at Los Alamos. This step was in part a response to the understanding that excellence and creativity in any endeavor are enhanced when the individuals involved are challenged by peers who also possess a recognized high level of skill and in part because it was felt by Edward Teller and others that Los Alamos Laboratory was not doing enough to develop a hydrogen device. The establishment of a second NEP design laboratory was also a risk-reduction step, broadening the portion of design space considered and reducing the risk of groupthink or blind spots among the designers.

The new Livermore laboratory immediately demonstrated that it was taking a fresh look at weapons design, as evidenced by the fact that its first few designs, tested in 1953 and 1954, failed. By 1955, Livermore designs began to succeed in tests and in meeting new warhead requirements from the U.S. military. The goal of exploring fresh approaches led to a distinct and different technical culture and approach at Livermore.

After the establishment of the Livermore laboratory in 1952, it was decided to also establish a branch of Sandia at Livermore. This site was set up to have an independent system design capability that would draw on the science and engineering technical base that already existed at Sandia in Albuquerque. The Sandia-Livermore site became operational in 1956.

Sandia National Laboratories (SNL) is responsible for the non-nuclear components of the warhead such as the radar and the neutron generator; the arming, fuzing, and firing system; and use controls, as well as integration of the entire package with the delivery system. Los Alamos National Laboratory (LANL) and Lawrence Livermore National Laboratory (LLNL) are responsible for the NEP. Because SNL components could be tested much more extensively, SNL is treated separately from LANL and LLNL in the discussion below.

TESTING, PEER REVIEW, AND DESIGN COMPETITION IN THE NNSA NATIONAL SECURITY LABORATORIES

Testing/Experimentation

Los Alamos National Laboratory and Lawrence Livermore National Laboratory

The definitive feedback about the quality and result of the designers' work during this period came from the direct validation enabled by nuclear explosion testing. According to the Comprehensive Nuclear-Test-Ban Treaty Organization (CTBTO),[1] from 1945 to 1992, the United States carried out 1,032 nuclear explosion tests, 815 of which took place underground, primarily at the Nevada Test Site.[2] The tests generally involved the NEP but not the entire warhead as an integrated system. In many cases, it is probably more accurate to call those tests "nuclear explosion experiments," since many new design concepts were explored, and a number did not perform as expected when actually tested. Both the successes and the failures helped the scientists better understand the interplay between design and functionality of the NEP. In addition, the nuclear explosion experiments provided data to help validate and improve the computational models of the warheads that were put into the nuclear stockpile. Those computational models could then be used to estimate weapon performance in conditions that differ from those represented in underground testing.

Sandia National Laboratories

As noted above, Sandia is responsible for the non-nuclear components of the warhead. As such, testing of Sandia components could occur more readily in aboveground facilities, and this testing was not substantially inhibited by the 1992 nuclear explosion testing moratorium. Indeed, even before the moratorium, Sandia had moved away from reliance on underground nuclear explosion tests.

---

[1] See http://www.ctbto.org/fileadmin/user_upload/pdf/Sipri_table12b.pdf.

[2] U.K. nuclear tests after 1962 were conducted jointly with the United States at the Nevada Test Site and are not included in this number.

Peer Review

Los Alamos National Laboratory and Lawrence Livermore National Laboratory

Each NEP laboratory did its best, using allocated resources, to develop and check its design, and the nuclear system test itself showed whether the developers were correct. Internal and external reviews of the science base, computer codes, and design work were conducted to ensure that the science base and the physics models in the code were the best that could be created with the knowledge available at that time. Reviews and extensive tests were also performed on electrical and mechanical systems that supported the functioning of the weapon. Interlaboratory peer review was less frequent and less formal than that which exists today, but it did occur on occasion, especially after nuclear test failures. In one famous case, a concept for a bomb-pumped x-ray laser that underwent development and preliminary testing by researchers at LLNL was shown by LANL to be infeasible.

Sandia National Laboratories

In the 1980s, Sandia initiated the independent Surety Assessment Center, which features a full- time team that conducts and contributes to the independent assessment of safety, use control, significant finding investigation (SFI) analysis and resolution, reliability analysis founded on certification and surveillance, and assurance of quality control. The team reported directly to the laboratory deputy director and director, which allowed maintaining significant independence from the design teams.

In the 1990s, Sandia relied on the independence between its New Mexico and California sites for effective independent technical peer review. Although the two sites were nominally part of the same laboratory, independence was gained through an organizational structure that separated nuclear weapons work at each site and geographically separated the nuclear weapons teams.[3]

Historically, Sandia has had a close relationship with the commercial electronics industry, which has stimulated intellectual cross-fertilization and also provided a kind of informal peer review of Sandia concepts and products.

---

[3] Gary Sanders, Sandia National Laboratories, "National Academy of Sciences Peer Review Study," presentation to the committee on June 10, 2014.

Design Competition

Los Alamos National Laboratory and Lawrence Livermore National Laboratory

Within Los Alamos itself, design competition existed from the start of the Manhattan Project and continued there until Lawrence Livermore was established. Scientists and engineers involved in the Manhattan Project pursued two design concepts in parallel, including all associated necessary functions spanning fissile material production, weapon design, and weapon production—in essence, this was the original nuclear weapon design competition.

The subsequent decades of the Cold War years were characterized by an intense design competition between the laboratories in Los Alamos and Albuquerque, New Mexico, and Livermore, California, which resulted in the robust and reliable U.S. nuclear deterrent. The impact of this independent, interlaboratory NEP design competition was profound. For example, significant reductions in size and weight of a thermonuclear weapon were achieved. The new designs could fit not only on intercontinental ballistic missiles (ICBMs), but even on missiles that could be carried on submarines. In an example from the later years of the Cold War, the design competition for the W88 warhead particularly emphasized the need to reduce the size of the warhead while reliably attaining a desired yield. Through this very active competition, both NEP laboratories achieved notable size reductions, again opening up new military options. In the later years of the Cold War, safety and security systems also benefited significantly from the differences in approaches that resulted from design competition, such as the development of a new, insensitive high-explosive system.

Sandia National Laboratories

During this era, the nation had two distinctly separate and competing nuclear weapon design capabilities: the Los Alamos/Sandia Albuquerque (LANL/SNLA) design collaboration team and the Lawrence Livermore/Sandia Livermore (LLNL/SNLL) design collaboration team. The Livermore branch of Sandia took on much of the culture of LLNL in that it took pride in designing non-nuclear systems for nuclear weapons that were different from those that its colleagues in Albuquerque were producing for Los Alamos while still relying on many component design teams (e.g., for the radar and neutron generator) in Albuquerque. Similar to the case of design competition between Los Alamos and Lawrence Livermore, design competition in the non-nuclear components and systems of the nuclear weapon produced significant advances in safety and security systems, such as the detonator safety technologies, that exist in the stockpile today. The Livermore site of Sandia continues to be active, but its work is now part of a distributed program across both the Albuquerque and the Livermore sites.[4]

OBSERVATIONS

More than 50 formal design competitions were held between the two NEP design laboratories during the Cold War period. These competitions resulted in the deployment of several new weapon types in the nation's stockpile and a stronger understanding of the relationship between different designs for NEPs and the consequent performance and reliability of a weapon type. When tests failed to meet design expectations, the results were shared among the three laboratories to identify and resolve design flaws.

It is important to note that these design competitions exercised all necessary aspects of the nuclear weapons complex, from design to engineering, fabrication, and testing of a prototype; each of these latter steps also provided feedback to all participants in the process, strengthening their judgment and insight. For the designs that were downselected for military systems, the production capability of the complex was also fully exercised. This work enabled the United States to meet the constantly evolving national security threat effectively throughout the Cold War era.

Based on discussions with laboratory experts at its meetings and on the accumulated experience of its members, the committee concluded that the following were key elements of the nuclear weapon design activities during the period of nuclear explosion testing (1945-1992):

- Scientific and technical staff of high competence and good judgment.
- Overlapping expertise between the two NEP design laboratories in Los Alamos and Livermore and between the two SNL sites, and the fact that the teams from the two locations used different technical approaches, which in combination made it possible to pursue genuine competition between two distinct teams with comparable competencies.
- A culture in each of the laboratories that encouraged extensive internal technical reviews to check the work being done.

---

[4] In some cases, Sandia took action to assign work to the Livermore branch that was previously conducted in Albuquerque because it was recognized that the volume of design work assigned to the Livermore branch at that time was not sufficient to maintain the design expertise.

- Vigorous competitions leading to successful weapon designs that were produced, tested, and fielded. As recounted in this chapter, design competitions led to significant tangible benefits both to the NEP and to system designs. These noteworthy successes led to the committee's strong endorsement of design competition, discussed later in this report.
- A strong sense of an international nuclear threat.
- Nuclear explosive testing, which provided definitive validation and feedback to the weapons designers and engineers.

While the last of these elements is not a viable option today, the others are largely available as contributors to the nuclear deterrent.

3

The Present: From 1992 Until Today

After the testing moratorium in 1992, nuclear explosion testing was no longer an option for the nuclear explosive package (NEP) design laboratories. The Stockpile Stewardship Program (SSP, originally called the Science-Based Stockpile Stewardship Program) was announced in 1995 by President Bill Clinton to sustain the U.S. nuclear weapons arsenal without that source of validation and feedback. The SSP is focused on improving the weapons science base and the computational and experimental capabilities in order to better ground the weapons systems in reliable scientific foundations and thus reduce the uncertainties associated with the inability to perform a full system test of the weapons. At the same time, and for the same reasons, the three laboratories developed and implemented over time an extensive intra- and interlaboratory technical review methodology to enhance their confidence in the stockpile stewardship work that dominates the mission. Approximations to design competitions were also implemented in an attempt to reestablish practices that had been so successful during the Cold War.

TESTING/EXPERIMENTATION

The Department of Energy's (DOE's) Stockpile Stewardship and Management Program[1] called for construction of a number of major experi-

[1] This May 1995 document responded to a Presidential Decision Directive and an act of Congress (P.L. 103-160). The Department of Energy was directed to "establish a steward-

mental facilities and advanced computing capabilities to strengthen the underlying scientific understanding in the nuclear weapons program in the absence of nuclear explosion testing.[2] Among the experimental facilities are the National Ignition Facility (NIF) at LLNL, the Dual-Axis Radiographic Hydrodynamic Test (DARHT) and the Los Alamos Neutron Scattering Center (LANSCE) at LANL, and the predecessors of the pulsed power Z machine at SNL, to name a few. On the computing side, the plan led to programs culminating in today's Advanced Simulation and Computing (ASC) Program.[3] These facilities and capabilities have improved our understanding of weapons physics, have enabled the performance of unique basic science experiments, and have attracted high-quality scientists and engineers to participate in the weapons program. The data generated are useful in peer reviews and for providing feedback to NEP designers, but these facilities and capabilities as a whole do not exercise the complete set of skills needed for actual design and engineering of an NEP.

Sandia National Laboratories

Tests of non-nuclear components and of subsystems and systems integrated with the delivery systems continued after the test ban. In the past and today, testing (with mock NEPs) of nuclear warheads integrated with their delivery systems is conducted for surveillance of stockpile systems and for certification of new design for Life-Extension Programs (LEPs).

System-level tests and reviews are used extensively in the SNL design process, design qualification, and surveillance. These include flight tests of warheads integrated with delivery systems and various environmental tests on full systems in the laboratory to qualify elements of the stockpile-to-target sequence.

Following the test ban, stockpile stewardship has relied more heavily on modeling and simulation (M&S), especially for NEP work. SNL's testing and experimentation continues to provide essential input for verification and validation of M&S at the component, subsystem, and system levels. The use of M&S in the non-nuclear design area has been driven by two factors: budget pressure to reduce costly testing (especially system tests) and the desire for better insight that comes from M&S used in concert with the traditional testing approach.

---

ship program to ensure the preservation of the core intellectual and technical competencies of the U.S. in nuclear weapons." http://fas.org/nuke/guide/usa/doctrine/doe/st01.htm.

[2] JASON, 1994, "Science Based Stockpile Stewardship," JSR-94-345.

[3] See, for example, http://www.nnsa.energy.gov/asc.

PEER REVIEW

Peer review and experimental validation are time-tested and widely respected mechanisms for reducing risk in research and development. This includes the risk that an unstated assumption or an ingrained practice may bias one's thinking or the outcome of an analysis—biases that might not otherwise be spotted in the absence of nuclear testing.

In this context, rigorous and effective peer review is needed to increase the degree of confidence the laboratories have in their scientific and technical work, thus contributing in a fundamental way to the laboratories' ability to perform annual assessments concerning the reliability of the nuclear weapons. This chapter discusses the mechanisms used today by the laboratories for peer review and documents the committee's view about how effective these reviews are.

At the same time that the laboratories have relied more on peer review, there have been fewer opportunities for design competition. This issue and these opportunities are also discussed in this chapter.

In general, the processes used today for peer review of nuclear weapon designs or development plans are more formal and less ad hoc than they were during the Cold War. Then, as now, all three laboratories conducted a large number of technical and programmatic reviews using reviewers primarily from their own institutions, as can be seen from the definitions of peer review that are provided in Chapter 1.

The committee asked each NNSA laboratory to estimate the number of nuclear weapons-related systems and science peer reviews that it participates in during a typical year. The laboratories were asked to classify these reviews into one of three categories: (1) reviews internal to the laboratory; (2) those involving external reviewers but still within the NNSA nuclear weapons complex (e.g., involving another laboratory or plant); and (3) those involving at least some reviewers external to the weapons complex, such as military customers, academics, or international organizations such as the Atomic Weapons Establishment of the United Kingdom (AWE).

The three NNSA laboratories used somewhat different criteria for rolling up their numbers in the different categories. For example, two of the laboratories had included work on foreign nuclear weapons assessments and nuclear forensics, while one had not. As a result, the committee agreed with the laboratories that it would not be useful to compare precise numbers across all three laboratories; rather, the committee draws the following semiquantitative conclusions:

- Each laboratory participates in some 800 nuclear weapons system and/or weapons science-related peer reviews in a given year.

- About half of those reviews are internal to the individual laboratories. These reviews are under their control, and they believe this to be about the right number for the work they do. The other half of these reviews involve some experts from one or both of the other NNSA laboratories and/or from outside the nuclear weapons complex.
- Substantial staff time and resources are involved in organizing, preparing for, conducting, and responding to these reviews.

Since the start of Stockpile Stewardship, the NEP design laboratories and SNL have evolved distinctly different approaches to peer review. Because of the NEP laboratories' reliance on simulations of complex physical processes, which are inherently imperfect, and SNL's ability to test a much larger portion of its work, it is reasonable for the three laboratories to have different approaches to peer review, as described here.

Los Alamos National Laboratory and Lawrence Livermore National Laboratory

LANL and LLNL rely a good deal on one another through interlaboratory peer reviews where the NEPs are concerned. There is no perfect model for NEPs, and experiments with nuclear materials tend to be expensive (and nuclear explosion testing experiments are not allowed at all). Therefore, computational simulations using different tools[4] are relied on extensively to predict NEP behavior. When teams with different perspectives use simulations based on different assumptions and methods and still obtain similar results, both laboratories tend to trust the scientific validity of those results. In contrast, if the two teams obtain disparate results, as happens sometimes, they work to resolve the differences, and that process leads to an improved understanding of how to model the NEPs.

At LLNL and LANL, about half of the reviews related to nuclear weapons or nuclear weapons science are totally internal. These peer reviews are controlled by the individual laboratories and the number is judged to be appropriate for the work they do. Of the remaining weapon-related reviews, about half are conducted with some participation external to the laboratory, and the other half are conducted with some participation from experts outside the NNSA weapons complex. Substantial effort is required to organize, prepare for, conduct, and respond to all of these peer reviews. In contrast to SNL, the committee did not hear about a formal

---

[4] Currently, the NEP laboratories use different codes and physics models, e.g., different plutonium equations of state.

written review process in the NEP design laboratories except in the case of INWAP reviews, red team reviews, reviews for the 6.x process (described below), and reviews to resolve SFIs (see section "Tri-Laboratory, Formalized, or Mandated Peer Reviews of Nuclear Warhead Systems," below).

With many staff and managers moving from one laboratory to another today, and with continued encouragement from DOE/NNSA and others for the two NEP laboratories to cooperate—including sharing codes and analysis practices—some observers have raised the concern that the laboratories' independence may be compromised by moving toward more commonality. The laboratory directors should monitor the composition of peer review panels to avoid this possibility.

Sandia National Laboratories

Although Sandia has an advantage over the NEP laboratories in that it is able to test its components and systems, peer review continues to be very important at SNL. One reason is the requirement to be right the first time with hardware introduced into the stockpile; another is the budget pressure to reduce the expense of testing, especially at the systems level.

Sandia has a more formalized and documented approach to peer review than do the NEP laboratories. There are two basic approaches. The first is the more formal interlaboratory peer reviews that are mandated by the *NNSA Development and Production Manual*[5] to occur at key milestones in a weapon or warhead acquisition process. For these reviews, one or more technically qualified individuals from LANL or LLNL may participate, if appropriate for the subject being reviewed. Participation by technically qualified personnel from AWE and other agencies is also allowed, if appropriate. For interlaboratory peer reviews that include production issues, participation by technically qualified personnel from the applicable production agency would be appropriate.

The other type of peer review at Sandia is internal reviews that are conducted to manage technical risk throughout a product's life cycle. The degree of rigor and the composition of the peer review panel are determined using a graded approach that depends on the degree of technical complexity and risk. The formal documentation of this procedure was finalized in September 2014.[6]

One requirement for these internal reviews is the inclusion of individuals with recent and relevant technical experience that is independent

[5] *NNSA Development and Production Manual*, Chapter 3.7, Inter-laboratory Peer Review Process.

[6] Sandia National Laboratories Product Realization Assurance Committee, NW Realize Product Procedure, RPP-12, Peer Reviews, September 29, 2014.

of the product being reviewed. Although SNL has no full-spectrum external peer organization in the sense that LANL and LLNL are peers of one another, there are times when the appropriate expertise and independence can be found in individuals from outside Sandia, resulting in peer reviews that include experts from other NNSA sites, industry, AWE, universities, and DOD.

SNL management has established within its organization the Independent Surety Assessment (ISA) team to perform surety-related assessments of the SNL systems in the stockpile.[7] The ISA team reports to the SNL laboratory director and is populated by laboratory technical experts who are assigned to ISA for as long as 3 years. It is separate organizationally from the activities it reviews. The ISA team has no members outside SNL.[8] SNL also conducts some reviews with all external members, reporting to NNSA as well as to the Sandia board's Mission Committee. The Mission Committee also reviews details of the design and engineering of nuclear weapons, including, for example, specification of tolerances and their impact on manufacturability.

Committee discussions with SNL presenters during the meeting of September 23, 2014 (see Appendix B), focused primarily on peer reviews at the component and subsystem levels. The committee's probing indicated that these reviews only occasionally involve technical experts from outside SNL, in contrast to SNL's guidance (adopted later that month; see footnote 24), which calls for broad participation by outside experts in peer reviews. This situation may be due in part to the fact that SNL does not have a counterpart in the United States with the full depth and breadth of expertise in non-nuclear components for nuclear weapons that could mirror the intimate interplay seen between LLNL and LANL with respect to the NEP.

Tri-Laboratory, Formalized, or Mandated Peer Reviews of Nuclear Warhead Systems

Reviews at each of the three laboratories are held with varying degrees of formality, ranging from fairly casual consultations with colleagues to carefully managed and process-driven reviews of high-stakes work. Interlaboratory peer review (IPR) involving all three laboratories is

[7] "Nuclear weapons surety refers to the materiel, personnel, and procedures that contribute to the security, safety, and reliability of nuclear weapons and to the assurance that there will be no nuclear weapon accidents, incidents, unauthorized weapon detonations, or degradation in performance at the target." Quoted from *The Nuclear Matters Handbook: Expanded Edition*, Nuclear Weapons Council, 2011, Washington, D.C., p. 24.

[8] Details about ISA are drawn from J.F. Nagel, "Organization 400 Independent Assessment," SNL presentation to the committee on September 23, 2014.

often conducted under a more formal process. For example, an independent review is required as part of the Annual Assessment Report (AAR) process for a particular weapon, as prescribed both by law and by directives of NNSA. Another important process in the AAR is the use of a red team, which was stipulated by law in 2003. These teams are formally IPR teams that report to the laboratory director and contain representatives from all three laboratories; their job is to vigorously challenge the assumptions and assessments made by a laboratory design team.

In 2000, the joint DOD-NNSA Nuclear Weapons Council (NWC) formulated a process for the life cycle of nuclear weapons that describes the evolution from concept to retirement in seven phases.[9] Currently the U.S. stockpile is largely in Phase 6 (production and maintenance), because present and future stockpile work will be focused on refurbishment[10] and maintenance and not on new weapons concepts. In 2004, NNSA formalized the process of preparing for and conducting LEPs for weapons, which is commonly called the "6.x process."[11]

NNSA's 6.x process subdivides Phase 6 into stages that partially map onto the original seven phases of weapon life cycles prescribed by the NWC. An IPR team, populated as needed by experts from the design laboratory not being reviewed and the Department of Defense,[12] is to be established in the 6.2 phase (which addresses the feasibility of various design options, the downselection to a provisional design, and a cost study). The IPR and design teams for a given weapon work together to prepare a plan and schedule for the needed peer reviews and their documentation.

A predecessor to the 6.x process—in that it used peer review as an input into planning and executing an LEP—was the dual revalidation of the W76 weapon in 1996 and 1997. The program included intensive and extensive peer review by separate and independent[13] LLNL/Sandia Livermore and LANL/Sandia Albuquerque teams to assess the W76 and all of its systems and components in order to identify those components and/or systems that needed to be modified and upgraded in the W76

[9] See Nuclear Weapons Council, "Procedural Guideline for the Phase 6.X Process," April 19, 2000, for detail on phases 6.1 through 6.6. The seven steps may be seen in DOD Instruction 5030.55, available at http://www.dtic.mil/whs/directives/corres/pdf/503055p.pdf.

[10] The 2000 NWC memo defines refurbishment as referring "to all nuclear weapon alterations and modifications to include life extension, modernization, and revised military requirements."

[11] See Chapter 3.7 in the latest version, 56XB, Rev. 2, 03-31-14; see also the presentation to the committee by Kevin Greenaugh of NNSA, June 10, 2014.

[12] Ibid.

[13] Sandia systems design teams were separate and independent but drew mainly from Albuquerque for their component groups.

LEP.[14] The two teams worked independently in a multiyear effort to assess and revalidate the ability of a specific weapon to meet its military requirements. Each team was required to evaluate the weapon's design, test history, and history of surveillance in depth, a process that called for true peer expertise. This intellectually and technically competitive effort was presented to the committee as an example of a peer review, not a design competition. In addition to being the first intensive post-moratorium peer review of a weapon system, this activity helped in setting up the steps in an LEP process and established lessons learned for future peer reviews. However, the dual revalidation process was not continued; it was replaced by smaller and less formal peer reviewer processes (e.g., INWAP, see below) that were perceived to be less cumbersome.

A somewhat similar review effort took place in 2001 when LLNL was given the task of refurbishing and rebaselining[15] the NEP for the W80 cruise missile warhead, a LANL design. This was the first example of a weapon whose continuing evaluation was assigned to the laboratory that had not designed it. While neither a peer review nor a competitive design, this project had aspects of both, because it was necessary for the LLNL team to thoroughly review the warhead's design as it planned its refurbishment—a process that might involve redesign and remanufacture of components of the NEP as well as redesign by SNL of components outside the NEP. Subsequently, the original NEP design laboratory (LANL) conducted several peer reviews of the LLNL work, and responsibility for the NEP was transferred to LLNL.

Another example of formalization of peer review is the Independent Nuclear Weapons Assessment Process (INWAP), which is part of the input to the annual assessment letter of the stockpile written by the LANL and LLNL directors. INWAP came about through the initiative of the NEP design laboratory directors and the NNSA in 2008, and it was subsequently authorized and funded by Congress. Under INWAP, each laboratory director is supported by a peer review team established at the other laboratory for an independent assessment of issues specified by the director regarding the safety, reliability, and effectiveness of each stockpile weapon for which his laboratory is responsible.[16] That peer review team assesses the given weapon—using the team's own approaches and computational models of that weapon and exercising their own tools to simulate weapon performance against the weapon's requirements—and

[14] The Navy Strategic Systems Programs (SSP) office had a considerable role in the final decision on the scope of the W76 LEP.

[15] Baselining a nuclear weapon involves constructing a physics and simulation model of the weapon that is capable of reproducing, to the needed accuracy, the nuclear and nonnuclear test history of that weapon.

[16] See LANL, LLNL, and SNL, "INWAP Implementation Plan," Initial Release 1.0, Effective February 16, 2010.

reports directly to the requesting director.[17] The requesting laboratory director then incorporates the independent assessment and the response from his own laboratory's design team as input to his annual assessment letter.

There are a number of positive aspects to the INWAP approach:

- It encourages each laboratory to regard the stockpile as a shared responsibility held jointly by all of the laboratories and plants, instead of as something in which one laboratory is responsible for a specific weapon. This builds trust and collaboration while at the same time valuing the differences in technical approaches, tools, and expertise.
- Because previous laboratory directors were personally involved in establishing INWAP, it is clear to all concerned that the heads of the NEP design laboratories value the benefit of alternative approaches and tools to inform their judgment. Since the results of the reviews are just for the use of the laboratory director, the stakes involved are lower since no funding decisions are involved, and any technical exchanges between the teams are much more likely to be constructive.
- If an issue arises with the NEP of a particular weapon, the two laboratories are well prepared to assist in addressing the issue, having already performed the necessary baseline work.
- As a result of the process whereby two technical teams fully share their approaches, expertise, and results, each team improves its own understanding of the weapon and areas that may be of future concern with respect to weapon reliability. The committee sees this as a strong incentive for bench scientists to engage in peer reviews.

According to some current and former laboratory directors, one important key to the success of INWAP is that it is carried out between the laboratories in such a manner that there is no winner and no loser, only an improved understanding of the stockpile. INWAP, which is strongly supported by the laboratories, is very beneficial because it accomplishes just that.

---

[17] This emphasis on using independent approaches and models stems from the recognition that no one model or simulation is a perfect proxy for reality. Good models provide insights, and examining a phenomenon through multiple models can provide a broader set of insights while reducing the risk of being misled by spurious model-dependent results. Multiple steps are needed to represent a model in a computer code, and the use of different approaches (e.g., different numerical algorithms), none of which are uniquely suited, similarly reduces the risk of being misled by a particular choice.

Other Types of Review

As noted above, the committee learned that all three laboratories conduct a large number of reviews (many hundreds) of nuclear components, technology, or systems in any given year. Most of the reviews require staff to invest a substantial amount of time in preparation, which must be weighed against the value gained from the review.

Many of these reviews are required by each laboratory's management and operating (M&O) contract and are initiated by the M&O contractor's governing board, laboratory management, or laboratory project/program managers. Review bodies initiated by, and reporting to, the M&O contractor board tend to be populated by a wide spectrum of people from outside the laboratory with backgrounds relevant to the scope of the laboratory's mission. Often they involve experts who work in the weapons program but who are not associated with the project being reviewed. In some instances, a few experts may come from other sites within the NNSA complex: they are generally included on an ad hoc basis when their expertise has been judged to be important to strengthen the review group. The frequency of these reviews typically ranges from quarterly to annually.

While these bodies may be asked to review selected laboratory capabilities, they are not nuclear weapons review panels. Rather, they are considered advisory boards or committees that assess the quality and direction of laboratory research; they may also advise on basic research priorities, the technical health of the laboratory, and overall laboratory (or laboratory division) strategic direction and planning. Within the definitions given in Chapter 1, assessments by these bodies would fall into the category of reviews by subject-matter experts; they are not peer reviews per se because most of the members are not weapons scientists or engineers and hence not "peers" in the strict sense of the word. However, the committee views these external boards as essential to the health of each laboratory and to the long-term viability of the national security effort within NNSA. Notwithstanding that benefit, because these reviews can require considerable preparation time and effort on the part of laboratory personnel that could otherwise be spent on core research and development, the committee counsels that these reviews be targeted thoughtfully and efficiently.

While not falling within the realm of the peer reviews that are the subject of this study, it is useful to recall those advisory panels that exist to provide input to each of the laboratories. For example, SNL has established the Nuclear Weapons External Advisory Board to give critical reviews and advice to Sandia; it reports to the SNL deputy director. It is composed of members of the Air Force, the Navy, the NEP laboratories, and the U.S. Strategic Command—STRATCOM. The NEP design laboratories have analogous advisory panels. Most of these advisory panels report to senior management at the relevant laboratory and meet once or twice a year.[18]

In addition to laboratory-conducted reviews, the committee notes that STRATCOM conducts an annual review of the status of the nuclear weapon systems deployed. This review is conducted by individuals retired from the weapon design laboratories and retired members of the DOD nuclear forces. Its outcome contributes to the Strategic Commander's input into the annual assessment letter and has historically identified areas of future research or heightened surveillance for the design laboratories. The committee understands this to be the only example of a standing technical review of a weapon or a weapon system and agrees that this typically amounts to a weapon system review. While this annual review does not delve into the technical depth the committee would associate with a true peer review, committee members associated with this process believe that the format of this review has been very valuable in ensuring the warheads are reliable.

Examples of Value Provided by Peer Review of Nuclear Warhead Work

In discussions during committee site visits at all three laboratories, examples were cited that highlighted the value of peer review. These examples showed how individuals outside the NNSA complex (e.g., from AWE, U.S. industry, or academia), and on occasion without nuclear weapon experience, provided valuable insight into a problem that had not been fully recognized by the design laboratory.

Some examples that were provided by the laboratories of the value of *external* peer review include the following:

- *Plutonium materials properties*. These include the equation of state (EOS) and kinetics properties (e.g., behavior under different strain rates and different timescales for different phenomena). Clear understanding of these properties of plutonium is critical to the analysis of weapons performance. However, the two NEP design laboratories use fundamentally different approaches to developing the EOS from very different perspectives. In recent years, intensive evaluation of how each laboratory's approach does, and does not, fit the data led to resolution of some long-standing differences. As a result, the two laboratories have attained a deeper

[18] Private communication between Gerry Sleefe of SNL and committee member David Overskei, October 9, 2014.

and joint understanding of the EOS for plutonium, decreasing uncertainties associated with NEP simulations.

- *Safety architecture signal in an arming, fuzing, and firing (AF&F) system.* A presentation to the committee by SNL[19] illustrated how a mathematical error had gone unrecognized by SNL staff, in spite of in-house reviews, until it was identified by an external peer reviewer from AWE who had not worked on weapons systems. The Air Force subsequently implemented changes to mitigate this problem.
- *Pit lifetimes and plutonium aging.* As plutonium ages, it could change in ways that would have important implications for the performance of pits and the long-term viability of the stockpile. Until fairly recently, LANL and LLNL had different views on the extent to which the aging of plutonium is a concern, and they were designing experiments to provide information on plutonium aging and its impact on predictions. An individual without weapons experience but with nuclear fuel experience suggested an approach that resulted in better understanding of plutonium aging as it relates to pit performance and lifetimes, and this method has been adopted by both laboratories.[20]

All three laboratories also cited examples in which their *internal* review processes caught significant design or production issues that were then rectified. In one example cited by SNL, Sandia recognized the need to seek additional options to reduce cost in the B61-12 LEP. Peer review teams were formed to focus on partial component reuse, and a new option was identified that saved over $1 billion.[21] Several additional examples involved deterioration of a system or component that was found during internal peer review. In selected cases, though, the concern was not completely addressed by the production team. In one such case, it was suggested that more rigor in the peer review process would likely have provided the impetus to push for a more robust solution, because the results of the peer review had not been solid enough to convince the production team to invest resources in a rectification.[22]

In addition to the above list, some examples were given in which peer

---

[19] Jeff Brewer, "Unique Signals Technical Basis Peer Review," SNL presentation to the committee on September 23, 2014.

[20] This example arose during general committee discussion with LLNL staff on November 14, 2014.

[21] Gary A. Sanders, Sandia National Laboratories, "National Academy of Sciences Study of Peer Review," presentation to the committee on June 10, 2014.

[22] Discussion following presentation by Steve Harris, "B61 Spin Rocket Motor Igniter Peer Review," September 23, 2014.

review identified issues and recommended actions, but the customer (Air Force or Navy) did not implement the mitigation approaches suggested, perhaps for reasons of cost or the impact on other DOD systems or different priorities.[23] In such cases, peer review still provides value by offering options that a decision maker can consider, and also by increasing the degree of confidence all involved have in the thoroughness with which weapons systems have been scrutinized.

It is worth noting that few, if any, of these issues would have been found or addressed by nuclear explosion testing. While testing results provided a crude form of validation (or demonstration of failure) of a particular design, teasing out the reasons for test successes or failures given all of the underlying variables, as well as predicting the future performance of aging weapons, has required an increased understanding of weapon physics and rigorous peer review of calculations and experimental results—both during the testing era and up to the present.

DESIGN COMPETITION

As noted in Chapter 1, there have been no full design competitions for new weapons in more than 20 years.

Los Alamos National Laboratory and Lawrence Livermore National Laboratory

Recent interlaboratory competitions capture some of the flavor of the design competitions of the Cold War years, but there has been no recent design competition at the NEP laboratories that—as often occurred during the testing era—culminated in the actual production of a prototype to verify that the weapon design was producible or viable.

One of the design studies was for a warhead to be included in a new Air Force cruise missile. It included an extensive look at the NEP, and it was decided that an LEP would be conducted on the existing W80 warhead to fill this need. However, that program has been postponed to the mid-2020s. Another example is the 120-day study on the 3 + 2 initiative.[24] The goal of this study was to look at the strategic missile warhead

---

[23] Discussion on January 22, 2015, with senior staffers from the Navy's Strategic Systems Program.

[24] This initiative will move the nation toward a stockpile consisting of three interoperable warheads deployed across the submarine-launched ballistic missile (SLBM) and the intercontinental ballistic missile (ICBM) legs of the Triad and two interoperable air-delivered warheads or bombs. See U.S. Department of Energy, 2014, *Fiscal Year 2015 Stockpile Stewardship and Management Plan, Report to Congress*, http://nnsa.energy.gov/sites/default/files/nnsa/04-14-inlinefiles/2014-04-11%20FY15SSMP_FINAL_4-10-2014.pdf.

stockpile as a whole and consider how the stockpile could be refurbished to provide three weapons for the ICBM/SLBM forces. The intent was to improve the safety and security and also to minimize the number of warheads that need to be kept in reserve as a hedge against a failure in the deployed force, because the warheads could be on either Navy or Air Force missiles. The idea was to provide a certain amount of interoperability and commonality of parts among the three stockpile weapons through redesign and reuse. Part of the initiative involved the competitive design of several interoperable warheads that could be used by both the Navy and Air Force legs of the Triad. However, this program was paused in 2013 for 5 years by the Air Force and NNSA, and in neither case did the studies go beyond computational simulations of the potential options.

In the Reliable Replacement Warhead (RRW) design study, both LLNL/SNL-California and LANL/SNL-New Mexico produced rather innovative RRW designs. The selection process had each team critically review the opposing team's design in a large meeting that included staff from DOD and NNSA in addition to the designers and their management. Although substantive improvements to both designs were discovered through this opposing team review, the fact that the results of the reviews were presented in such a forum, and somewhat early in the process, created deep-seated negative feelings on the part of the two NEP laboratories and mistrust of NNSA that still exists. These unfortunate aspects of the manner in which the RRW competition was conducted were compounded by the fact that the program was canceled by Congress before either design had been validated by completing engineering and manufacturing a prototype—essential elements of any successful design competition. Specifically, in order to adequately exercise their design skills, designers must "close the loop" and, at the very least, receive feedback from the real world about whether their design is practical and can be manufactured.

Sandia National Laboratories

Sandia designers of non-nuclear components, subsystems, and systems integrated with the delivery system have more opportunities to exercise their skills today than do designers at the NEP laboratories. In addition to addressing the problem of deterioration, designers must replace components when obsolete technology is no longer amenable to remanufacture. Extensive LEPs for the W76 and B61 have also enabled Sandia designers to exercise their skills. The committee notes, however, that in these LEPs, the SNL designers were constrained to meet the interface requirements of the old NEPs, which were not changed. Thus, these cases do not stimulate the kind of innovation and creativity provided by a "clean slate" design competition.

CONCLUDING COMMENTS

Based on its extensive experience with nuclear weapons science and technology, the committee concludes that in the absence of nuclear testing, strong peer review and design competition create a higher level of confidence in the nation's stockpile than could be generated without them. The ultimate measure of success of peer review and design competition will be a sustained, competent, and creative workforce capable of responding to emerging challenges. The judgment of that workforce will be informed, in turn, by extensive and deep engagement with the scientists and engineers in an environment that fosters and encourages free questioning and checking of one another's work.

4

The Future: Responding to Evolving Challenges

Because the nuclear deterrent remains a cornerstone of U.S. national defense policy, assuring the quality of the technical work that supports the nation's stockpile is at least as important now as it has been since the introduction of nuclear weapons. However, owing to the U.S. adherence to the moratorium on nuclear weapons testing, the three NNSA laboratories must retain the design and engineering capability needed to maintain the nuclear stockpile without designing and testing new weapons. Maintaining the stockpile under these constraints requires a deep understanding of how weapons' surety and performance may be affected by aging or the substitution of components. At the same time, the laboratories must develop and exercise new skills to address evolving stockpile needs. For example, it would not be surprising if the nation were to someday need newly designed and engineered components to address some fundamental safety or reliability risk to the nuclear stockpile. The laboratories would not only have to design and engineer the new components but would also have to assess their effect on the overall weapons in which they are placed. Moreover, as other nations pursue new designs or strategies that could constitute serious threat evolutions, the United States could find itself in a precarious security situation were it not to maintain nuclear weapon design, development, and production skills to address such evolving demands.

During the Cold War era, the nation depended on an active program of designing and testing nuclear weapons because the threat from other nations was changing rapidly. Following the collapse of the Soviet Union, and perhaps even in the later years of the Cold War, the nuclear threat landscape appeared more static, and the United States stabilized and reduced its stockpile. After the cessation of nuclear explosion testing, the United States and its allies put their main efforts into maintaining the existing stockpile.

Today the nation is again facing evolving threats as Russia and China modernize their nuclear stockpiles and their doctrines for use. Several countries are contemplating the development of nuclear capabilities, terrorism has spread across the world, and technology advances are accelerating.

NNSA is entrusted with the responsibility of ensuring unimpeachable confidence in the nation's nuclear warheads through its nuclear complex, consisting of the three national security laboratories, the test site, and the production plants. Today the technical challenges to carrying out this mission in the global environment described above are unprecedented:

- For the near term, the average age of the warheads in the stockpile will continue to climb, increasing the challenge of surveillance, meant to assure the physical state of the weapons, and for the annual stockpile assessments.
- For the medium term, Life-Extension Programs (LEPs) for the warheads in all the enduring stockpile systems are planned over the next several decades. In an LEP, components that are known to suffer from deterioration or obsolescence within the time frame considered in the LEP are replaced; other components that may degrade more slowly with time or fail abruptly after a time longer than that examined may or may not be replaced. As a result, the "aging clock" of these latter components continues to tick. In some LEPs, components may need to be changed to improve the safety and security of the warhead. Assessment of the performance of changed components is a significant challenge, especially because the skills associated with designing and developing nuclear explosive packages (NEPs)—and, in the process, strengthening understanding of the linkages between design and performance—have not been thoroughly exercised in the complex for more than 20 years.
- For the longer term, the nation must have in place the capability to anticipate and respond to the potential new threats the country could face in the future. To attempt this with a future workforce without validated experience in weapon design and development would be very risky.

These technical challenges all require a science and engineering enterprise of high quality and technical staff of high competence and good judgment. Peer review and design competition contribute essentially to that quality.

CONCLUSIONS AND RECOMMENDATIONS

Ultimately, the nuclear weapons program can only be effectively maintained (and a rigorous system of peer review preserved) if many of the nation's best scientists and engineers choose to commit their time and talents to the NNSA laboratories. The most effective way of encouraging this is to maintain technical vitality at the laboratories through state-of-the-art research, exciting work, challenge, and reward. In support of their core mission—to sustain the nation's nuclear deterrent—the laboratories should seek to develop and support the people who execute that mission. While a comprehensive treatment of this challenge is beyond the scope of the current report, it has been the subject of numerous recent reports dealing with the quality of science and engineering at the laboratories[1] and governance of the laboratories.[2]

Below, the tasks that constitute the committee's charge are taken from Chapter 1, repeated one by one, and set in italics. Each task is followed by the committee's relevant conclusions and recommendations.

Assess the quality and effectiveness of peer review of designs, development plans, engineering and scientific activities, and priorities related to both nuclear and non-nuclear aspects of nuclear weapons.

The examples cited in Chapter 3 illustrate how technical programs within the NNSA national security laboratories have benefited from peer review of several kinds. While the approaches used by all three laboratories have proven successful in the main—providing high-quality, effective peer reviews of designs, development plans, and engineering and scientific activities—the committee finds that there are areas that need to be strengthened in these approaches in order to improve the assurance of stockpile reliability and surety going forward.

---

[1] National Research Council (NRC), 2012, *Managing for High-Quality Science and Engineering at the NNSA National Security Laboratories*, The National Academies Press, Washington, D.C.; NRC, 2013, *The Quality of Science and Engineering at the NNSA National Security Laboratories*, The National Academies Press, Washington, D.C.

[2] Congressional Advisory Panel on the Governance of the Nuclear Security Enterprise, 2014, *A New Foundation for the Nuclear Enterprise*, November; NRC, 2015, *Aligning the Governance Structure of the NNSA Laboratories to Meet 21st Century National Security Challenges*, The National Academies Press, Washington, D.C.

In particular, some of the case studies presented to the committee illustrate that relying too heavily on in-house peers and experts can limit the value of reviews because a review group composed solely of insiders may be so close to the technology and program under review that it fails to recognize subtle weaknesses in components, systems, or methods of analysis. In contrast, the engagement of peers with different experience bases, perspectives, and technical skill sets can strengthen a review process and avoid groupthink. The committee recognizes that involving some experts from outside the laboratory can add some costs and perhaps entail a learning curve, but it believes the value gained from outside perspectives offsets those downsides. The case studies presented to the committee also demonstrated the value of having documented frameworks for reviews. Such frameworks codify best practices and in essence capture the insights gained during the days of nuclear testing, when laboratory staff saw the many ways in which nature confounded the best predictions of any single team of scientists and engineers. Codifying the peer review approaches developed in recent years with the insights from the days of nuclear testing can be very beneficial to future generations of weapons designers after the current generation has retired.

Conclusion 1.1: In the main, peer review processes used by all three NNSA laboratories are healthy and robust, providing benefits such as increasing confidence in weapon assessment and certification, improving our understanding of weapons physics, addressing weapon aging issues, and identifying lower-cost approaches to Life-Extension Programs.

Assess incentives for effective peer review.

In a world where the nuclear threat is evolving, peer review becomes even more important to NNSA and its laboratories (in the absence of testing of the NEP and with pressure on budgets for testing of non-nuclear subsystems and systems) for addressing technical challenges encountered in carrying out their mission. The primary value and incentive for peer review is that it reduces risk—of overlooking a technical option, of relying on suboptimal data or methods, or of simply making a mistake. In essence, peer review improves the quality of the work. As noted in Chapter 3, a process by which two technical teams compare their approaches, expertise, and results helps each team improve its own understanding of a weapon and its safety, security, and effectiveness. That knowledge is essential and intrinsically valued by staff at all levels in the laboratories. All three laboratories have an ingrained culture that sets a high standard for quality; this gives staff members an incentive to use peer review.

Furthermore, peer review is clearly valued by laboratory management. As an example, the NEP design laboratory directors were directly involved in establishing and advocating for INWAP, which has improved understanding of the stockpile and reinforces confidence in the warhead annual assessments. The value senior leadership places on peer review provides a strong incentive for peer review among the middle management and staff. The committee notes that independent peer reviews are regularly used by laboratory directors to assess early design feasibility studies as well as to review surveillance and SFI results, even though such reviews are not mandated and may expend their limited resources. In summary, the committee concludes the following:

Conclusion 1.2: Incentives for peer review at the NNSA laboratories are abundantly evident.

- **Peer review reduces the risk of overlooking a technical option, of relying on suboptimal data or methods, or of simply making an embarrassing mistake, and thereby supports the ingrained culture at all three laboratories, which sets a high standard for quality.**
- **Peer review is visibly valued by laboratory management.**

Assess the potential effectiveness, efficiency, and cost of alternative methods of conducting peer review and design competition related to both nuclear and non-nuclear aspects of nuclear weapons, as compared to current methods.

Conclusion 1.3: SNL and the NEP design laboratories (LANL and LLNL) have taken somewhat different approaches to peer review, owing in large part to SNL's ability to test non-nuclear components and systems.

- **With only archival nuclear explosion test data available, LANL and LLNL rely on vigorous, deep-dive reviews by truly competitive peers and other subject-matter experts to critique the results of calculations and subcritical experiments relating to NEP performance.**
- **With testing data available to verify the performance of components and systems and to validate modeling and simulation tools, peer review at SNL is driven more by the need to assure cost-effective performance of stockpile hardware under all anticipated conditions and by budget pressures to reduce the number of expensive tests.**

Conclusion 1.4: All three NNSA laboratories have opportunities to improve their processes for peer review:

- **With the exception of major reviews associated with the Annual Assessment Report or Life-Extension Programs, LLNL and LANL lack written guidance for conducting peer reviews to determine in general when a review is needed, how the review is to be conducted, who should participate in the review, or how to address review findings.**
- **SNL has developed useful written guidance for conducting peer reviews; however, during its visit to SNL and its probing of the presentations made, the committee determined that SNL could profitably make greater use of outside experts in its peer reviews, as called for in its written guidance.**

Recommendation 1: The nuclear weapon laboratories should improve their peer review processes in the following ways:

- **Los Alamos National Laboratory and Lawrence Livermore National Laboratory should ensure they have short, written guidance for a graded approach to peer review, the rigor of which is appropriate to the stage of work and range of technical activities being reviewed.**
- **Sandia National Laboratories should strengthen and broaden its use of outside experts on its peer review teams, as articulated in written guidance that Sandia recently finalized.**

Conclusion 2: The innovations produced by design competitions during the Cold War, as well as the increased confidence in the safety and reliability of stockpile weapons resulting from current assessment processes such as the Independent Nuclear Weapons Assessment Process (INWAP), illustrate the value of having independent teams, using different approaches and methods, addressing common problems.

Recommendation 2: Los Alamos National Laboratory and Lawrence Livermore National Laboratory should continue to maintain independent design capabilities, using different approaches and methods, to enable independent peer review of critical technical issues. Sandia National Laboratories should likewise carry out, for high-priority issues, competitive designs with independent teams that use different approaches, followed by peer reviews of components, subsystems, and full systems as discussed in Recommendation 1.

Recommendation 1, which calls for the NEP laboratories to develop additional peer review guidance and for SNL to ensure that outside experts are fully utilized in peer reviews, could entail some additional costs, though these should be minor. The call of Recommendation 2 for the maintenance of capabilities to conduct independent analyses does involve additional costs, but the benefits are likely to outweigh these.

Assess known instances where current peer review practices and design competition succeeded or failed in finding problems or potential problems.

There are not many examples where peer review has resulted in major course corrections; some of the more important ones are listed in Chapter 3. These include an improved understanding of the plutonium equation of state and of how plutonium's characteristics change as it ages.

The limited number of examples in part reflects the fact that the research and development teams at the laboratories are really quite good and that they get it right most of the time. The other factor that is missed entirely by making lists of successes and failures is that a large part of the benefit of the review comes in the preparation for the review. The people to be reviewed have to get their thoughts in order about what they are doing and how to present what they have done and what they need to do next. This may be the most important benefit. The (expert) designers who are being reviewed by (expert) peer reviewers are more likely to themselves realize they have made a mistake than are the reviewers, especially since most mistakes are subtle and are found only when designers are recalculating everything that is being reviewed and staring at the calculation frames or movies that do not seem quite right. The committee looked for instances in which peer review failed to find problems, but it did not find any clear examples of this.

The three laboratories currently carry out high-quality, effective peer reviews of designs, development plans, and engineering and scientific activities. However, in the areas of design studies and design competition—which often include interlaboratory technical reviews—the committee found situations in need of improvement (see Conclusions 2 and 3 of Chapter 3, repeated below).

The Reliable Replacement Warhead (RRW) design study, discussed in Chapter 3, was one such particularly difficult example from which many lessons should be learned. The technical review format with broad participation early in the process engendered deep-seated and negative concerns between the two laboratories and a mistrust with NNSA that still exists. While the RRW was successful in motivating the nuclear weapon design staff and in generating unique designs, the competition process had faults, and it did not offer any validation of the design because there was no opportunity to explore the viability of the "winning" design. It is not the right model to follow for future efforts to exercise the laboratories' design capabilities.

Conclusion 3: Although the RRW design study succeeded in producing innovative weapon designs by the competing teams, its value was reduced because technical experts from the competing laboratories were not given the opportunity to critique one another's ideas through interlaboratory peer review or to address criticisms at the science and engineering level before the final designs were formally presented to NNSA and potential end users.

Senior staffers from both LANL and LLNL told the committee[3] that they support the concept of true design competitions as a necessary means of maintaining the laboratories' capabilities in nuclear weapons design. But they emphasized the need for constructively managed competitions that are initialized with well-elucidated guidelines and a clearly envisioned outcome. In addition, they agreed that paper studies alone are not enough to challenge and maintain the skills of the weapon designers.

Recommendation 3: To guide future design studies and design competitions, the National Nuclear Security Administration (NNSA) should provide a formal written statement articulating the design requirements and objectives, along with the selection criteria, in advance of any authorized work. NNSA should also ensure that interlaboratory peer review takes place and that competitors have an opportunity to address criticisms at the science and engineering level before the results are formally presented to stakeholders outside NNSA.

Recommendation 3 calls for a process change in future design competitions that would have no effect on costs.

As the threat faced by them continues to evolve, the United States and its allies will face new challenges. Since the end of the Cold War, the perception has been that the threats facing the United States have not required any basic change in the capability of the nuclear deterrent. Over the past two decades, Congress has restricted new NEP design studies and DOD has not required any fundamentally new warhead designs, nor have there been any of the associated design competitions that were so

[3] Jas Mercer-Smith, during committee discussions on September 24, 2014, and Charles Verdon, during committee discussions on November 14, 2014.

successful and valuable to the health of the complex during the Cold War. Because of this lack of new NEP design work, essential capabilities have not been exercised in a generation and are at risk. The committee's most significant concern is that the capability for a full, integrated, end-to-end design, including of the integrated NEP—from the design concept for a different device to address a threat through the production of an engineering prototype at a level of confidence that it could, in principle, be considered for the stockpile—is not being exercised. The approved programs for the future (e.g., the W76-1 and the B61-12 LEPs) have not included the full physics design challenges because these programs are largely just a rebuild of the original NEP design with some changes of materials where necessary. The few "design competitions" since 1992 have been largely extensive design studies that did not move beyond the study phase to the later phases of full engineering development and prototype hardware and an evaluation of whether the design might be acceptable for the stockpile. The fraction of the NEP laboratories' science and engineering personnel with hands-on experience in nuclear weapons design or testing continues to decrease and will reach zero in the next decade or so. Once this experience is lost, it could seriously compromise the nation's defensive posture and will be difficult to reestablish.

To keep this from happening, the NNSA complex needs a way to exercise the full suite of nuclear weapons design, development, and production capabilities. A true design competition would exercise the full spectrum of skills and activities needed to produce a weapon that qualifies for inclusion in the stockpile. Such capabilities might be needed, for example, if evolving military requirements require an adjustment to an NEP.

Conclusion 4: In contrast to the robust state of peer review at the NNSA laboratories, the state of design competition is not robust.

- **There have been no full NEP design competitions since the 1992 nuclear explosion testing moratorium. Recent design studies have been good analysis and modeling exercises, but they did not result in the actual engineering and fabrication of components and systems; thus, they did not exercise the complete set of skills required in the NNSA complex to design nuclear weapons that would be an effective deterrent, nor was the credibility of any design assessed by fabricating a device or by non-nuclear testing.**
- **At SNL, the need to continually replace aging or obsolete non-nuclear components in stockpile weapons, as well as the large Life-Extension Programs for the W76 and B61, have indeed exercised designers' skills. However, these exercises do not stimulate**

the full creativity and innovation that result from a true blank slate design competition that includes engineering and building a prototype.

Looking to the future, maintaining nuclear weapon design skills at the NEP laboratories—as well as production skills within the NNSA complex—is essential to achieve three objectives:

1. Maintaining a credible nuclear deterrent workforce that is capable of designing and building weapons to meet evolving threats;
2. Understanding the status and direction of foreign nuclear weapon programs and thus strengthening the nonproliferation regime;
3. Determining the best and most cost-effective approaches to resolving problems that arise during stockpile weapon surveillance and Life-Extension Programs.

Recommendation 4: In order to exercise the full set of design skills necessary for an effective nuclear deterrent, the National Nuclear Security Administration should develop and propose the first in what the committee envisions as a series of design competitions that include designing, engineering, building, and non-nuclear testing of a prototype. The non-nuclear components produced by Sandia should be integrated into the design and fabrication of the prototype. This should be done with the clear understanding that this prototype would not enter the stockpile.

These design competitions should be of modest cost and managed so as to impact neither the cost of nor the schedule for delivering LEPs or the 3 + 2 program plan. The projects should be full design competitions that involve LANL, LLNL, SNL, and representatives from the plants and an applicable military service. Such design competitions should be initiated periodically (perhaps once every 5 years) to allow learning from mistakes and for the continuous development of judgment and skills of the nuclear weapon enterprise workforce. This recommendation is not unprecedented. In 2005, an Advisory Board (SEAB) Task Force of the Secretary of Energy recommended that a new version of the RRW, "incorporating new design concepts and surety features, be initiated on planned 5-year cycles."[4]

The committee recognizes that if the 3 + 2 program was implemented,

[4] U.S. Department of Energy. Secretary of Energy Advisory Board, Nuclear Weapons Complex Infrastructure Task Force, 2005, *Recommendations for the Nuclear Weapons Complex of the Future,* July 13, p. 13.

BOX 4.1
Prototype Nuclear Device Characteristics

The prototype product of the design competitions of Recommendation 4 should have the following characteristics:

- The prototype design and production should exercise the full range of skills in the NNSA complex needed to produce a new weapon.
- The design should be such that the consensus of the design community is that it could be certified in a manner consistent with the nuclear testing moratorium—that is, it should be close enough to a vetted design to permit that consensus.
- The prototype NEP should be fully integrated with all Sandia components needed for a warhead. Prototypes of new Sandia components should be designed and produced in parallel.
- It should be a "nuclear device," not a warhead. That is, stockpile-to-target-type scenarios should be considered via simulation or testing, but there should be no expectation of flight testing.

there would be some warhead redesign work associated with the interoperable warheads (e.g., if conventional high explosive was to be replaced by insensitive high explosive).[5] In addition, prototype interoperable warheads could be built. However, because of the constraints associated with the requirements for component reuse and compatibility with existing delivery systems, the committee's view is that the 3 + 2 program does not involve fundamentally new designs and therefore does not exercise the same NEP design skills as the "clean slate" design competitions recommended here. To the extent that the 3 + 2 program does turn out to involve extensive warhead redesign, it could fulfill in part the purpose of Recommendation 4.

The committee realizes that Recommendation 4 will be controversial, particularly its call for NNSA to hold periodic competitions at its laboratories that produce a prototype nuclear weapon. The committee's concept of the characteristics of such a prototype is laid out in Box 4.1.

Recommendation 4 might be seen by some critics as promoting an aggressive posture that would put the United States in a position to manufacture new nuclear weapons quickly and thus fuel a new global nuclear arms race. These same arguments were made in the vigorous

[5] JASON, 2015, "Technical Considerations for the Evolving U.S. Nuclear-Weapons Stockpile Executive Summary," JSR14-Task-006E, January.

debate over the wisdom of the RRW program and its potential effects on nonproliferation efforts.[6] And, as noted earlier, a similar recommendation was made by the 2005 SEAB Task Force.

Congress has authorized NNSA to "develop and carry out a plan for the national security laboratories and nuclear weapons production facilities to design and build prototypes of nuclear weapons to further intelligence estimates with respect to foreign nuclear weapons activities."[7] More recently, Congress called for "the directors of the national security laboratories [to] jointly develop a multiyear plan to design and build prototypes of nuclear weapons to further intelligence estimates with respect to foreign nuclear weapons activities and capabilities."[8] Recommendation 4, which is aimed at the preservation of nuclear weapon design capabilities at the NNSA laboratories, is consistent with the spirit of these authorizations.

Recommendation 4 calls for alternative design competitions that would be much more effective than the recent design studies, but it would entail costs. The committee's view is that the laboratory staff activities that would take place during its recommended competitions would be focused applications of science and engineering skills that should happen anyway under the laboratories' science campaign (~$412 million in FY2015) and engineering campaign (~$136 million in FY2015), so that the incremental cost of the competitions should be modest, while the benefit to the nation would be immense.

Roughly speaking, the committee imagines a design competition as involving a few dozen laboratory staff members, with a larger number in the first year of each competition, plus some prototype development and experiments up to and including hydrodynamic tests.[9] These parameters suggest a scale for the endeavor that the committee deems appropriate.

Assess how peer review practices related to both nuclear and non-nuclear aspects of nuclear weapons should be adjusted as the three NNSA laboratories transition to a broader national security missions.

Over the past 60 years, the mission of the NNSA laboratories has

[6] Jonathan Medalia, Congressional Research Service, 2009, *The Reliable Replacement Warhead Program: Background and Current Developments*, July 27.

[7] Section 3115 of the 2013 National Defense Authorization Act (P.L. 112-239).

[8] Section 3111 of the 2015 National Defense Authorization Act (P.L. 113-291).

[9] "In a hydrodynamic test, inert material (e.g., ^{238}U or a simulant for plutonium) is imploded to determine how well the high-explosive system functions. The testing program for an unboosted implosion device primarily ensures that the hydrodynamic behavior of the implosion (particularly of a hollow pit) is correct." Federation of American Scientists, http://fas.org/nuke/intro/nuke/test.htm.

evolved from an exclusive focus on designing, engineering, testing, and maintaining nuclear weapons to a broader and more diverse mission of advancing national security generally. In response to funding uncertainty and the needs of other government agencies, the laboratories have worked to build a set of clients beyond DOE. This broadened funding base supports the core capabilities needed to perform the laboratories' primary nuclear security mission. From modest beginnings in the 1960s, when DOD asked one of the laboratories to develop sensors for a specific application in the Vietnam War, what was formerly called Work for Others (WFO) and is now termed a Strategic Partnership Project (SPP) has grown to become a significant portion of the laboratories' technical portfolios. In FY2013, for instance, the NNSA weapons complex received $1.656 billion in research funding from other federal agencies,[10] accounting for 36 percent of SNL's budget, 18 percent of LLNL's, and 9 percent of LANL's.

The largest sponsors of SPP at the NNSA laboratories are DOD and the Intelligence Community. Other sponsors include the Department of Homeland Security, the National Institute of Standards and Technology, the Food and Drug Administration, the Centers for Disease Control and Prevention, and the National Aeronautics and Space Administration. In 2008, Secretary of Energy Samuel Bodman formally articulated a vision for the future of the NNSA laboratories as national security laboratories charged with conducting research and development to address a range of national security threats facing the nation.[11]

Several recent reports have noted the benefits that SPP brings to the nuclear weapons mission of the laboratories. For example, SPP provides challenging problems to the laboratory scientists that help attract, develop, and retain key personnel.[12] Furthermore, as illustrated by several presentations during the committee's meeting at SNL, in many areas there are direct correlations between technical projects pursued in the context of SPP and nuclear weapons work (e.g., on radar or other sensor technologies). This results in an intellectual ferment that enriches the weapons program.

In the present context of assessing the effects of the evolving national security mission of the NNSA laboratories on peer review in the nuclear weapons program, the committee concludes that SPP will expand the base of technical experts available for peer review by involving (1) expert personnel inside the laboratories but outside the direct weapons programs

---

[10] NRC, 2014, *Aligning the Governance Structure of the NNSA Laboratories to Meet 21st Century National Security Needs*, The National Academies Press, Washington, D.C.

[11] "Transforming the Nuclear Weapons Complex into a National Security Enterprise," signed on June 19, 2008.

[12] NRC, 2012, *Managing for High-Quality Science and Engineering at the NNSA National Security Laboratories*, The National Academies Press, Washington, D.C.

and (2) SPP customers in DOD, the Intelligence Community, and their industry partners, who will add substantially to the pool of qualified peer reviewers for the weapons program. These new sources of expertise could help broaden and diversify laboratory peer reviews, as called for in Recommendation 1.

SUMMARY COMMENTS

Implementation of the above four recommendations would help ensure that the most important asset—a competent workforce with demonstrated skills and judgment—is being developed and maintained and that all stakeholders (including our adversaries) have confidence in that workforce. There is no better way to learn and to develop judgment than to evaluate and test one's ideas and to understand how implementation compares with expectations. The judgment of this workforce is the basis for all stakeholders' confidence in the nuclear deterrent and the NNSA laboratories' ability to respond to evolving national threats.

Appendixes

A

Biographical Sketches of Committee Members

PAUL S. PEERCY (*Co-Chair*) is a professor emeritus at the University of Wisconsin, Madison, where until 2013 he had served as dean of the College of Engineering. He came to the university in September 1999 from SEMI/SEMATECH, where he had been president since 1995. Before that, he was director of the Microelectronics and Photonics R&D Core Competency at Sandia National Laboratories. His research interests include phase transitions in solids, ferroelectricity, Raman and Brillouin scattering studies of solids, ion–solid interactions, laser-induced phase transformations, microelectronics and photonics, and solid state devices. He is the author or coauthor of more than 185 technical papers and holds two patents. Dr. Peercy is a distinguished member of the Tau Beta Pi Engineering Honor Society; a fellow of the American Association for the Advancement of Science (AAAS), the American Physical Society (APS), and the Institute of Electrical and Electronics Engineers (IEEE); and a member of the National Academy of Engineering (NAE). He received his Ph.D. in physics from the University of Wisconsin, Madison, in 1966. He was a member of the 2012-2013 NRC Committee to Review the Quality of the Management and of the Science and Engineering Research at the Department of Energy National Security Laboratories—Phase II.

JILL P. DAHLBURG (*Co-Chair*) is superintendent of the Space Science Division at the Naval Research Laboratory (NRL) and a member of the Senior Executive Service since December 2007. Dr. Dahlburg served as NRL senior scientist for science applications from June 2003 to December

2007. From 2001 to mid-2003, she left NRL to work for General Atomics as the director of the Division of Inertial Fusion Technology and codirector of the Theory and Computing Center. In 2000, she served as head of the NRL Tactical Electronic Warfare Division Distributed Sensor Technology Office, where she was co-principal investigator for the first year of development of the small, expendable unmanned aerial vehicle Dragon Eye, which saw active duty in Iraq. Dr. Dahlburg holds a B.A. (1978) from St. John's College in Annapolis and an M.S. in physics (1980) and a Ph.D. in theoretical physics (1985) from the College of William and Mary. A fellow of the APS, she has served as chair of the APS Division of Plasma Physics, chair of the APS Topical Group on Energy Research and Applications, chair of the APS Panel on Public Affairs, and chair of the APS Mid-Atlantic Section. She also served as a member of the Lawrence Livermore National Laboratory (LLNL) Defense and Nuclear Technologies Director's Review Committee (2001-2007), chair of the Department of Energy (DOE) Advanced Scientific Computing Advisory Committee (2005-2007), and chair of the Department of the Navy's Space Experiments Review Board (2006-present). She was a member of the 2012-2013 NRC Committee to Review the Quality of the Management and of the Science and Engineering Research at the Department of Energy National Security Laboratories—Phase II.

JOHN F. AHEARNE is executive director emeritus of Sigma Xi, the Scientific Research Society; emeritus director of the Sigma Xi Ethics Program; and an adjunct professor of engineering at Duke University. Before working at Sigma Xi, Dr. Ahearne served as vice president and senior fellow at Resources for the Future and as commissioner and chair of the U.S. Nuclear Regulatory Commission. A member of the NAE, he has been in the White House Energy Office and served as deputy assistant secretary of energy. He also worked on weapons systems analysis, force structure, and personnel policy as deputy and principal deputy assistant secretary of defense. Serving in the U.S. Air Force, he worked on nuclear weapons effects and taught at the Air Force Academy. Dr. Ahearne's research interests include risk analysis, risk communication, energy analysis, reactor safety, radioactive waste, nuclear weapons, materials disposition, science policy, and environmental management. He earned his Ph.D. in physics from Princeton University in 1966. He was a member of the 2012-2013 NRC Committee to Review the Quality of the Management and of the Science and Engineering Research at the Department of Energy National Security Laboratories—Phase II.

MICHAEL R. ANASTASIO was, until June 2011, director of Los Alamos National Laboratory (LANL) and president of Los Alamos National Security, LLC, the company that manages and operates LANL. He was previously director of LLNL from 2002 until 2006. Dr. Anastasio holds a bachelor's degree in physics from Johns Hopkins University and M.S. and Ph.D. degrees in theoretical nuclear physics from the State University of New York at Stony Brook. His career at LLNL began in 1980 as a physicist in B-Division, one of the two nuclear weapons design divisions. He participated in the development of the W87, W84, and B83 warheads and with 10 nuclear tests; he was project physicist on four of these tests. In 1991, he was made B-Division leader and program manager responsible for primary weapons design. From 1996 to 2001, he served as associate director for Defense and Nuclear Technologies, responsible for all activities in the laboratory's nuclear weapons program. In that capacity, he was instrumental in the development and execution of the national Stockpile Stewardship Program. From 2001 to 2002, Dr. Anastasio served as LLNL's deputy director for strategic operations. He received the DOE's Weapons Recognition of Excellence Award in 1990 and the DOE/NNSA Gold Medal in 2011. He is a member of the U.S. Strategic Command's Strategic Advisory Group, the Defense Science Board, the Draper Laboratory Corporation, and the boards of Los Alamos National Security, LLC, and Lawrence Livermore National Security, LLC. He was a member of the Congressional Advisory Panel on the Governance of the Nuclear Security Enterprise.

CHRISTINA A. BACK is the advanced nuclear materials leader at General Atomics and an experimental physicist with expertise in radiation in high-energy-density plasmas. She received her B.S. in physics from Yale University in 1984 and her Ph.D. in plasma physics from the University of Florida in 1989. After a 2-year postdoc with the Centre National de la Recherche Scientifique at the Ecole Polytechnique in France, she spent 13 years at LLNL in the Inertial Confinement Fusion (ICF) and HEDS programs, specializing in the study of radiation transport and spectroscopy. In addition to serving on committees of the NRC and the APS, she is an APS fellow and has served that society as a general councilor. Currently, she is target production coordinator and radiation physics manager at General Atomics, where she is responsible for identifying new opportunities for target fabrication advances in high-energy-density physics (HEDP) as well as conducting research to develop novel radiation sources for lithographic and other applications. With her knowledge of experimental methods and target requirements, she interfaces closely with colleagues in the national and international ICF and HEDP research programs. She was a member of the 2012-2013 NRC Committee to Review the Quality of the Management and of the Science and Engineering Research at the Department of Energy National Security Laboratories—Phase II.

JOHN M. CORNWALL is professor emeritus at the University of California, Los Angeles (UCLA), where he conducts theoretical research on elementary particles. He is an inventor of the pinch technique and co-author of the 2011 book *The Pinch Technique and Applications to Non-Abelian Gauge Theories*. A member of JASON, Professor Cornwall has a long history of interactions with, and service to, the NNSA laboratories, including as long-time chair of the LLNL nuclear weapons review committee and with the NNSA Predictive Science Panel.

PAUL A. FLEURY is the Frederick W. Beinecke Professor of Engineering and Applied Physics and professor of physics at Yale University, where he also serves as director of the Yale Institute for Nanoscience and Quantum Engineering. He received his Ph.D. from MIT in 1965 and was employed at Bell Laboratories from 1965 until 1996, spending a portion of that period as a vice president at Sandia National Laboratories. Following his time at Sandia, Professor Fleury served as dean of the School of Engineering at the University of New Mexico (1996-2000) and then dean of engineering at Yale (2000-2007). His research has been in experimental condensed matter physics and materials science. A member of the NAE and the National Academy of Sciences (NAS), he is also a member of the American Academy of Arts and Sciences and a fellow of the AAAS. He received the Michaelson Morley prize (1985) and the Frank Isakson prize for optical effects in solids (1992) from the APS.

DAVID HAMMER is the J. Carlton Ward Professor of Nuclear Energy Engineering and professor of electrical and computer engineering at Cornell University. He has been a Cornell faculty member since 1977. Dr. Hammer worked at NRL from 1969 to 1976, was a visiting associate professor (part-time) at the University of Maryland from 1973 to 1976, and was an associate professor at UCLA in 1977. He spent sabbatical leaves from Cornell in 1983-1984, 1991, and 2004 as a visiting senior fellow at Imperial College, London, in 1998 at Applied Materials, Inc.; and in 2011 at the Paris Observatory in France. His research interests are in high-energy-density plasmas, especially as it relates to inertial confinement fusion.

CHERRY A. MURRAY is dean of Harvard University's School of Engineering and Applied Sciences; John A. and Elizabeth S. Armstrong Professor of Engineering and Applied Sciences; and professor of physics. Dr. Murray served as principal associate director for science and technology at LLNL from 2004 until 2009 and was APS president in 2009. Before joining LLNL, she was senior vice president of physical sciences and wireless research at Bell Laboratories Research, where she worked for 27 years. A member of both NAS and NAE, she has served on more than 100 national and international scientific advisory committees, governing boards, and NRC committees and as a member of the National Commission on the BP Deepwater Horizon Oil Spill and Offshore Drilling. Until January 2014, she chaired the NRC's Division on Engineering and Physical Sciences. As an experimentalist, Dr. Murray is known for her scientific accomplishments in condensed matter and surface physics. She received her B.S. in 1973 and her Ph.D. in physics in 1978 from MIT. She has published more than 70 papers in peer-reviewed journals and holds two patents in near-field optical data storage and optical display technology.

DAVID OVERSKEI is the founder and president of Decision Factors, Inc., which has provided strategic and management consulting on complex problems since 2004. Dr. Overskei has led internationally recognized fundamental and applied research teams in academia and private industry in topics related to defense, national security, energy technologies, medical systems, optical and wireless communications, and advanced software for command and control. He chaired the Nuclear Weapons Complex Infrastructure Task Force of the Secretary of Energy's Advisory Board. The task force was responsible for a major analysis of the facilities, personnel, and management skills required through 2030 to sustain a reliable and safe nuclear deterrent in a cost-effective manner. He has performed numerous studies for DOE on problems related to managing new technology programs, and he routinely advises the Department of Defense and U.S. allies on questions related to nuclear warhead maintenance and supply needs. He has served for the past 6 years on the Programme Advisory Committee for the Atomic Weapons Establishment in the U.K., and he served on the recent NRC study on the Assessment of the Governance Structure of the NNSA National Security Laboratories.

K. LEE PEDDICORD is director of the Texas Engineering Experiment Station, senior associate dean for research, and professor of nuclear engineering at Texas A&M University. His areas of research include behavior of nuclear fuels, reactor systems and design, fissile materials disposition, MOX fuels, Generation IV nuclear power systems, nuclear-generated hydrogen, the hydrogen economy, and the nuclear workforce. He received his Ph.D. in nuclear engineering from the University of Illinois in 1972. Professor Peddicord was a member of the 2012-2013 NRC Committee to Review the Quality of the Management and of the Science and Engineering Research at the Department of Energy National Security Laboratories—Phase II.

ROBERT SELDEN is a private consultant in defense science and research management. He retired in 1993 as an associate director at LANL. His career in the nuclear weapons laboratories began at LLNL in the 1960s when he was one of the two participants in the Nth Country Experiment to design a nuclear explosive from unclassified information. After moving to LANL in 1979, he served as the division leader of the Applied Theoretical Physics Division, as associate director for theoretical and computational physics, and as the first director of the Los Alamos Center for National Security Studies. Dr. Selden served as the chief scientist of the U.S. Air Force from 1988 to 1991, when he received the Air Force Association's Theodore von Karman Award for outstanding contributions to defense science and technology. He has been a member of the Strategic Advisory Group to the Commander of the United States Strategic Command since 1995. Since 2003 he has served as chairman of the Advisory Group's Stockpile Assessment Team, which has the responsibility for conducting a detailed annual review of the U.S. nuclear weapon stockpile. He also is currently a member of the Joint Advisory Committee on Nuclear Surety to the Secretaries of Defense and Energy, and he served on the recent NRC study on the Assessment of the Governance Structure of the NNSA National Security Laboratories. He was a member of the Air Force Scientific Advisory Board from 1984 to 2005. Dr. Selden received his B.A. degree from Pomona College, Claremont, California, in 1958 and his Ph.D. degree in physics from the University of Wisconsin in 1964.

STEVEN J. ZINKLE is the Governor's Chair Professor in the University of Tennessee's Nuclear Engineering Department, a position he took in 2013 after a long career at Oak Ridge National Laboratory (ORNL). Dr. Zinkle received a B.S. degree in nuclear engineering, M.S. degrees in materials science and nuclear engineering, and a Ph.D. in nuclear engineering from the University of Wisconsin, Madison. His specialties are the physical metallurgy of structural materials; the effects of ion and neutron irradiation on the microstructure, physical properties, and mechanical properties of metals and ceramics; transmission electron microscopy; and fusion and space fission reactor materials studies. Dr. Zinkle worked in the ORNL Metals and Ceramics Division since 1985, rising to become an ORNL corporate fellow, director of the Materials Science and Technology Division, and chief scientist for the Nuclear Science and Engineering Directorate. A member of the NAE, he is the author or co-author of more than 250 peer-reviewed scientific articles and is a fellow of seven professional societies, including APS, the Materials Research Society, the American Nuclear Society and the Minerals, Metals and Materials Society.

B

Information-Gathering Meetings of the Committee

MEETING 1
JUNE 10-11, 2014
NATIONAL ACADEMIES' KECK CENTER, WASHINGTON, D.C.

June 10, 2014

CLOSED SESSION

8:30 - 9:30 am Closed Committee Discussions

OPEN SESSION

9:30 Break

9:45 Discussion with NNSA Senior Staff

Dimitri Kusnezov, Kathleen Alexander, Robert Hanrahan, and Kevin Greenaugh

- NNSA expectations from the study
- Discussion of the statement of task, including the final item: "such other matters related to peer review and design competition related to nuclear weapons as the NNSA Administrator considers appropriate"

- High-level view of peer review and design competition at the laboratories

11:15 Discussion of study charge with congressional staff members

Drew Walter, Strategic Forces Subcommittee, House Armed Services Committee

- Clarify charge
- Discuss desired outcomes from the study

12:15 pm Working Lunch

1:00 Reflections on Peer Review and Design Competition: History, Current Practices, and Future Opportunities

Speakers *Gary Sanders, Sandia National Laboratories*
Michael Dunning, Lawrence Livermore National Laboratory
Michael Bernardin, Los Alamos National Laboratory

3:00 Break

3:15 Open Discussion about What Information to Request and What Questions to Discuss During Site Visits to the Three NNSA Laboratories

4:15 Discussion of Peer Review in Other Settings for Specialized R&D

John Lyons, former director of the National Institute of Standards and Technology (NIST) and of the Army Research Laboratory (ARL); chair of the NRC's Laboratory Assessments Board

5:15 Adjourn

CLOSED SESSION

6:00 - 8:00

June 11, 2014

CLOSED SESSION

8:30 am - 2:00 pm Closed Committee Discussions

MEETING 2
SEPTEMBER 23-25, 2014
SANDIA NATIONAL LABORATORIES, ALBUQUERQUE, N.M., AND LOS ALAMOS NATIONAL LABORATORY, LOS ALAMOS, N.M.

September 23, 2014, at SNL

DATA-GATHERING SESSION: NOT OPEN TO THE PUBLIC

8:00 am	Introductions and Logistics	*Ron Hartwig*
8:15	Welcome by Executive Management	*Paul Hommert*
8:30	Overview of Peer Review and Cost Competitiveness at Sandia • Motivation for Peer Review • Requirements and Principles for Sandia Peer Review • Evolution of Sandia Peer Review Process • Framework for Our Technical Basis and Reviews • Cost Competitiveness Approaches and Examples • Introduction to Case Studies	*Gary Sanders*
9:45	Break	
10:00	Requirements Review Case Studies • Unique Signals Technical Basis Peer Review • B61 Spin Rocket Motor Igniter Peer Review • W76-1 Inter-Lab Peer Review	 *Jeff Brewer* *Steve Harris* *Brent Blankenship*
11:00	Design Review Case Studies • W88 ALT 370 Peer Review	 *Celeste Drewien*

	• Explosive Component Peer Reviews	*Everett Hafenrichter*
11:50	Working Lunch—Discussion Continues	
12:35 pm	Qualification Review Case Studies	
	• Cougar ASIC Peer Review	*Perry Molley*
	• Qualification Alternatives to Sandia Pulse Reactor	*Charles Barbour*
	• Radar Peer Reviews	*John Moser*
	• B61-12 Peer Review	*Curt Nilsen*
2:00	Manufacturing Review Case Studies	
	• W87 Neutron Generator Peer Reviews	*Carla Busick*
	• Packaging Advisory Board	*Mike Kelly*
2:45	Break	
3:00	Life Cycle Independent Assessment Processes	*J.F. Nagel*
	• Org 400 Independent Assessment	
	• Red Team Process	
4:00	Sandia Summary	*Gary Sanders*
4:15	Wrap-up/Concluding Discussion	
5:00	Adjourn	

September 24, 2014, at LANL

DATA-GATHERING SESSION: NOT OPEN TO THE PUBLIC

8:00 am	Welcome, Security Review	
8:10	Overview Remarks of NAS Co-Chairs	*Jill Dahlburg, Paul Peercy*
8:15	Overview of Peer Review and Design Competition	*Robert Webster*

8:30	Design Competition and Stockpile Assessments • Nuclear Testing Era (50 minutes) • Post Nuclear Testing Era (40 minutes)	 *Michael Bernardin* *John Scott*
10:00	Break	
10:15	Pit Certification and Peer Review	*Maurice Sheppard*
11:00	SFIs—Connections to Design and Peer Review	*Patrick Garcia*
11:45	Discussion	
12:15 pm	Working Lunch—Discussion Continues	
1:15	Competition of Ideas and Role of Peer Review in Weapon Science	*Mark Chadwick*
2:00	Competition of Ideas and Role of Peer Review in Weapon Code Development	*Mathew Bement*
2:45	Break	
3:00	HPRF Design Competition • RRW Peer Review • Weapon System Margin • Second 120-Day Study	*Jas Mercer-Smith*
4:00	Future Design and Certification Challenges	*Charles Nakhleh*
4:30	Discussion	
5:15	Adjourn to Working Dinner	
5:45 - 8:30	Working Dinner to Continue Discussions with Senior LANL Staff Members	

END OF DATA-GATHERING SESSION: NOT OPEN TO THE PUBLIC

September 25, 2014, at LANL

CLOSED COMMITTEE MEETING: MEMBERS AND NRC STAFF ONLY

8:00 am	Discussions and Development of Draft Materials
2:00 pm	Adjourn

MEETING 3
NOVEMBER 14-15, 2014
LAWRENCE LIVERMORE NATIONAL LABORATORY, LIVERMORE, CALIF.

November 14, 2014

DATA-GATHERING SESSION: NOT OPEN TO THE PUBLIC

8:00 am	Welcome, Security and Safety Comments	*Michael Dunning*
8:05	Introductory Remarks by NAS, Introductions	*Dahlburg, Peercy*
8:20	Welcome by LLNL Senior Management	*Charles Verdon*
8:45	Peer Review and Competition in the Russian Laboratories	*Siegfried Hecker*
9:30	Historic Process-Motivated Peer Review and Design Competition • Dual Revalidation	*Joseph Bauer*
10:00	Break	
10:15	Historic Process-Motivated Peer Review and Design Competition (cont.) • W87 Life-Extension Program	*K. Henry O'Brien*
11:15	Ongoing Process-Motivated Peer Review • Annual Assessment Review of the Stockpile • Intelligence Assessments	*Susan Taylor* *Nils Carlson*

12:15 pm	Working Lunch	
1:15	Issue-Motivated Peer Review	
	• Cushion	*Cynthia Nitta*
	• W88	*Peter Raboin*
	• Pit Lifetimes/Plutonium Aging	*Patrick Allen*
2:45	Break	
3:00	Science-Motivated Peer Review	
	• National Ignition Facility/ National Ignition Campaign	*Michael J. Edwards*
	• Predictive Science Panel	*Michael Zika*
	• High-Z Experiments	*Arthanasios Arsenlis*
4:30	Future Design and Certification Challenges	*Des Pilkington*
5:00	Plenary Discussion	
5:30	Transfer to Dinner	
6:15 - 8:30	Working Dinner to Continue Discussions with Senior LLNL Staff Members	

END OF DATA-GATHERING SESSION: NOT OPEN TO THE PUBLIC

November 15, 2014

CLOSED COMMITTEE MEETING: MEMBERS AND NRC STAFF ONLY

8:00 am	Discussions and Development of Draft Materials	
2:00 pm	Adjourn	

SMALL-GROUP MEETING TO OBTAIN BRITISH INPUT JANUARY 21, 2015 NATIONAL ACADEMIES' KECK CENTER, WASHINGTON, D.C.

6:00 pm	Introductions and General Discussion	
6:30	Overview of Peer Review in the Atomic Weapons Establishment	*Graeme Nicholson, A.W.E.*

7:30 Dinner and Continue Discussion

8:00 Discussion of Committee Questions about Peer Review in the British Nuclear Weapons Enterprise — *David Overskei*

9:00 Adjourn

Attendance: Committee members Jill Dahlburg, Paul Peercy, and David Overskei
Committee staff Scott Weidman
Guests:
Graeme Nicholson, Atomic Weapons Establishment
Craig Shobrook, Strategic Weapons Project Team, U.K. Ministry of Defence
David Holder, Ministry of Defence, British Embassy, Washington, D.C.

SMALL-GROUP MEETING TO OBTAIN CUSTOMER INPUT JANUARY 22, 2015 RONALD REAGAN BUILDING, WASHINGTON, D.C.

12:00 pm Introductions

12:05 Comments on NNSA Laboratories' Peer Review and Design Competitions from Senior Staff of the U.S. Navy's Strategic Systems Programs

12:50 Wrap-up Discussion

1:00 Adjourn

Attendance: Committee members Paul Peercy and David Overskei
Committee staff Scott Weidman
Guests:
John Hill, Navy Strategic Systems Programs
Barry Hannah, Navy Strategic Systems Programs

C

Authorizing Language for the Study

Public Law No: 112-239, National Defense Authorization Act for Fiscal Year 2013, Sec. 3144:

SEC. 3144. NATIONAL ACADEMY OF SCIENCES STUDY ON PEER REVIEW AND DESIGN COMPETITION RELATED TO NUCLEAR WEAPONS.

(a) STUDY.—Not later than 60 days after the date of the enactment of this Act, the Administrator for Nuclear Security shall enter into an agreement with the National Academy of Sciences to conduct a study of peer review and design competition related to nuclear weapons.

(b) ELEMENTS.—The study required by subsection (a) shall include an assessment of—

(1) the quality and effectiveness of peer review of designs, development plans, engineering and scientific activities, and priorities related to both nuclear and non-nuclear aspects of nuclear weapons;

(2) incentives for effective peer review;

(3) the potential effectiveness, efficiency, and cost of alternative methods of conducting peer review and design competition related to both nuclear and non-nuclear aspects of nuclear weapons, as compared to current methods;

(4) the known instances where current peer review practices and design competition succeeded or failed to find problems or potential problems; and

(5) such other matters related to peer review and design competition related to nuclear weapons as the Administrator considers appropriate.

(c) COOPERATION AND ACCESS TO INFORMATION AND PERSONNEL.—
The Administrator shall ensure that the National Academy of Sciences receives full and timely cooperation, including full access to information and personnel, from the National Nuclear Security Administration and the management and operating contractors of the Administration for the purposes of conducting the study under subsection (a).
(d) REPORT.—
(1) IN GENERAL.—The National Academy of Sciences shall submit to the Administrator a report containing the results of the study conducted under subsection (a) and any recommendations resulting from the study.
(2) SUBMITTAL TO CONGRESS.—Not later than September 30, 2014, the Administrator shall submit to the Committees on Armed Services of the House of Representatives and the Senate the report submitted under paragraph (1) and any comments or recommendations of the Administrator with respect to the report.
(3) FORM.—The report submitted under paragraph (1) shall be in unclassified form, but may include a classified annex.

Following the first meeting of the study, discussions regarding element (5) of the study charge led to it being replaced with the wording shown in Chapter 1: "how peer review practices related to both nuclear and non-nuclear aspects of nuclear weapons should be adjusted as the three NNSA laboratories transition to a broader national security mission."

D

Acronyms

AAR	Annual Assessment Report
AF&F	arming, fuzing, and firing
AWE	Atomic Weapons Establishment of the United Kingdom
DOD	Department of Defense
DOE	Department of Energy
EOS	equation of state
ICBM	intercontinental ballistic missile
INWAP	Independent Nuclear Weapons Assessment Process
IPR	interlaboratory peer review
ISA	Independent Surety Assessment at SNL
LANL	Los Alamos National Laboratory
LEP	Life-Extension Program
LLNL	Lawrence Livermore National Laboratory
M&O	management and operating
M&S	modeling and simulation
NEP	nuclear explosive package
NNSA	National Nuclear Security Administration
NRC	National Research Council

NWC	Nuclear Weapons Council
Pu	plutonium
R&D	research and development
RRW	Reliable Replacement Warhead
S&E	science and engineering
SFI	significant finding investigation
SLBM	submarine-launched ballistic missile
SNL	Sandia National Laboratories
SNLA	Albuquerque branch of SNL
SNLL	Livermore branch of SNL
SPP	Strategic Partnership Project
SSP	Stockpile Stewardship Program
WFO	Work for Others